Environmental Science and Engineering

The ultimate goal of this series is to contribute to the protection of our environment, which calls for both profound research and the ongoing development of solutions and measurements by experts in the field. Accordingly, the series promotes not only a deeper understanding of environmental processes and the evaluation of management strategies, but also design and technology aimed at improving environmental quality. Books focusing on the former are published in the subseries Environmental Science, those focusing on the latter in the subseries Environmental Engineering.

Gehan Mohamed El Zokm ·
Mona Mohamed Ismail · Mariam Helmy Mostafa

Rethinking Seaweeds via Applications and Technologies

Composition, Toxicology, Economy and Sustainability

Springer

Gehan Mohamed El Zokm
National Institute of Oceanography
and Fisheries (NIOF)
Cairo, Egypt

Mona Mohamed Ismail
National Institute of Oceanography
and Fisheries (NIOF)
Cairo, Egypt

Mariam Helmy Mostafa
Faculty of Pharmacy
Arab Academy for Science, Technology,
and Maritime Transport
Alexandria, Egypt

Faculty of Pharmacy
Alexandria University
Alexandria, Egypt

ISSN 1863-5520 ISSN 1863-5539 (electronic)
Environmental Science and Engineering
ISBN 978-3-032-14955-8 ISBN 978-3-032-14956-5 (eBook)
https://doi.org/10.1007/978-3-032-14956-5

This Springer imprint is published by the registered company Springer Nature Switzerland AG
The registered company address is: Gewerbestrasse 11, 6330 Cham, Switzerland

Preface

Seaweeds, which are multicellular algae, serve as the foundation of marine food webs that thrive in intertidal zones. They can be divided into three groups based on their photosynthetic pigments: Chlorophyceae, Ochrophyceae, and Rhodophyceae. The highest nutritional density and best source of micronutrients are found in green seaweeds. Various factors, including biological, physical, and chemical ones, significantly influence the dispersal of seaweed. Metabolites, which can be thought of as biological systems' response reactions to genetic or environmental changes, are what define seaweeds. Their biochemical composition primarily supports their nutritional and therapeutic uses. The metabolites of a biological system can be comprehensively analyzed using an approach called omics, which allows for the simultaneous detection of hundreds of compounds. Omics research focused on seaweeds has the potential to rapidly expand our understanding of molecular biological processes and contribute to the resolution of today's most pressing environmental and socioeconomic issues by utilizing a range of computational and informatics frameworks. Additionally, seaweeds have various environmental and economic applications. Seaweeds play an important role in the ecosystem as primary producers, protecting shorelines, purifying water, and reducing the effects of rising ocean acidity and falling oxygen levels. Because of their bioactive content, seaweeds are becoming a competitive alternative to synthetic substances from an economic standpoint. They are used in a variety of industries, including biotechnology, nutrition, medicine, cosmetics, and coloring. They are recommended as a low-calorie functional food since they are a good source of protein with specific amino acid and fatty acid content. Seaweeds are a valuable resource for medical and biofuel production due to their ability to store carbohydrates as metabolic reserves. They have anticoagulant, antiviral, antioxidant, anti-inflammatory, and anticancer properties. Seaweeds can boost agricultural productivity and serve as versatile bioinoculants that are suitable for organic farming. In recent years, the process of extracting phytochemicals from seaweeds has gained popularity. Nevertheless, to improve the extraction efficiency and bioavailability of biocompounds present in algae, new technologies such as green approaches in conjunction with conventional extraction methods are required. The bioavailability and extraction efficiency of biocompounds found in seaweeds require the use of modern technologies, such as green techniques combined with traditional extraction techniques. To sustainably extract the most advantageous compounds from seaweed, a number of eco-friendly extraction methods have been employed. While there are numerous sophisticated and developed green extraction technologies, the most widely used is solid-liquid extraction (SLE). These technologies differ primarily in terms of recovery rate, time, volume, temperature, pH, ratio, and solvent type. When selecting an environmentally friendly extraction process, the main consideration is whether the recovered bioactive component is as effective and efficient as possible. The seaweed industry has grown quickly and is now a key area of the world economy. By enhancing aquaculture and facilitating the "Blue Carbon" storage of CO_2 to counteract climate change, marine macroalgae can help achieve several Sustainable Development Goals (SDGs), including SDG 13 (Climate Action) and SDG 14 (Life Below Water). SDGs 1 (No Poverty), 2 (Zero Hunger), 8 (Decent Work and Economic Growth), and 12 (Responsible Consumption and Production) all support the growth of the seaweed industry, and UN SDG 7 states that seaweed is a biomass feedstock for biofuels, reducing reliance on burning fossil fuels. The use of recyclable materials for farming operations and renewable energy sources for hatchery electricity should be given priority for improving the sustainability of seaweed products. These strategies have been identified as crucial for reducing agricultural emissions.

Cairo, Egypt
2025

Gehan Mohamed El Zokm
Mona Mohamed Ismail
Mariam Helmy Mostafa

Competing Interests The authors have no competing interests to declare that are relevant to the content of this manuscript.

Contents

About the Authors

Gehan Mohamed El Zokm is a professor at the Marine Chemistry Department and the head of the Marine Environment Division. She is the leader of the halogen, sulfide modeling, and surfactant groups at the National Institute of Oceanography and Fisheries, Mediterranean Sea Branch in Egypt. She got her B.Sc., M.Sc., and Ph.D. in the Chemistry Department at the Faculty of Science of Alexandria University, Egypt, both of which concerned heavy metal chemistry (focusing on preconcentration and fractionation tools). Her research interests are in analytical, instrumental, and environmental chemistry. Statistical analyses were the most prominent features of her research, which was conducted using the most recent software packages. Her papers discuss the topics of halogen chemistry, water quality, sulfur-sulfide chemistry, eutrophication, and the responses of marine biota to pollutants. Her researches also focus on the mobility and fluxes of various parameters within the marine environment. Extraction of natural products with many applications and nutritional studies is her recent project. Her research is also focused on water treatment and monitoring of pollutants (heavy metals, nutrients, phenols, hydrocarbons, pesticides, and surfactants). She applied factorial design to water treatment experiments. She is interested in the ecotoxicity of pollutants on Egyptian coasts. She reviews many papers for important publishers (Elsevier, Springer, and Taylor and Francis). She headed and participated in several projects related to the marine environment. She supervised many master's and Ph.D. students, participated in conferences and workshops, and had several activities. She heads the Master's and Doctoral Thesis Registration Committee at the National Institute of Oceanography and Fisheries. She has 65 research papers and five books with an h-index of 20.

Mona Mohamed Ismail is a professor of Taxonomy and Biodiversity of Aquatic Biota, specializing in Phycology at the National Institute of Oceanography and Fisheries in Alexandria, Egypt. She is the head of the Biodiversity and Taxonomy of Aquatic Biota Laboratory. She got her Ph.D. in 2010 and her M.Sc. in 2005 from the Faculty of Science, Tanta University, Egypt. Dr. Ismail's research focuses on key scientific issues of biodiversity, algal taxonomy, classification, identification, biodiversity, and biotechnology. She has published over 65 papers in international journals with an h-index of 18. Dr. Ismail is an editorial board member of some scientific journals in the environment and related sciences and is a reviewer for more than 80 international scientific journals following Elsevier, Springer, MPDI and Taylor & Francis. Since 2014, she headed and participated in several national projects and continues to supervise master's and doctoral theses. Teaching of different phycology and microbiology courses for undergraduate and postgraduate students for the Faculties of Science and Education of Tanta and Kafe El Sheekh universities in Egypt. Sharing in Organizing of different Workshops and conferences belonging to NIOF, Egypt.

Mariam Helmy Mostafa is a teaching assistant in pharmacognosy at the College of Pharmacy, Arab Academy for Science, and Technology, and Maritime Transport, class of January 2024. Now pursuing her master's degree at the College of Pharmacy, Alexandria University. Her research interests focus on pharmacognosy, phytochemistry, phytotherapy, functional foods, sustainable research techniques, nutrition, flavour development, and herbal cosmetics, where she has published two scientific papers in the pharmacognosy field and established her H-index Score. Beyond academia, Mariam is a member of the EU-JEEL Connect Network, a network of the European Union under the EU Neighbours South Programme. Through this role, she succeeded in founding an initiative called Bridging Cultures through Natural Products.

Abbreviations

AAS	Amino Acid Score
ABA	Abscisic Acid
ACE	Angiotensin I Converting Enzyme
ACh	Acetylcholine
AD	Alzheimer's Disease
AGEs	Advanced Glycation End Products
Ala	Alanine
Arg	Arginine
ASE	Accelerated Solvent Extraction
Asp	Aspartic Acid
AT	Abu Talat
BHA	Hydroxyl Anisole
BHT	Butylated Hydroxyl Toluene
BOD	Biological Oxygen Demand
CCA	Crustose Coralline Algae
CDR	Carbon Dioxide Removal
CNCs	Cellulose Nanocrystals
COD	Chemical Oxygen Demand
DHA	Docosahexaenoic Acid
DOC	Dissolved Organic Carbon
DPP-4	Dipeptidyl Peptidase 4
DVM	Diel Vertical Migration
EAAs	Essential Amino Acids
EH	Eastern Harbor
EM	El Mex Bay
EPA	Eicosapentaenoic Acid
EROEI	Energy Returned on Energy Invested
ESE	Enhanced Solvent Extraction
F6P	Fructose-6-Phosphate
FAO	Food and Agriculture Organization
F-AuNPs	Fucoidan-Functionalized Gold Nanoparticles

G6P	Glucose-6-Phosphate
GC	Gas Chromatography
Glu	Glutamic acid
Gly	Glycine
GRFT	Griffithsin
His	Histidine
HRAP	High-Rate Algal Pond
HVED	High voltage electrical discharge
IAA	Indole Acetic Acid
IBA	Indole Butyric Acid
IBV	Infectious Bronchitis Virus
JEV	Japanese Encephalitis Virus
LC	Liquid Chromatography
LCAs	Life Cycle Assessments
Leu	Leucine
Lie	Isoleucine
LLE	Liquid-Liquid Extraction
Lys	Lysine
MAAs	Mycosporine-like Amino Acid
MAE	Microwave-Assisted Extraction
MAPK	Mitogin-Activated Protein Kinases
MB	Methylene Blue
MEF	Moderate Electric Field
Met	Methionine
MPAs	Management of Marine Protected Areas
MS	Mass spectrometry
NK	Natural killer cells
NMF	Natural Moisturizing Factor
NMR	Nuclear Magnetic Resonance Spectroscopy
NSAIDS	Nonsteroidal Anti-Inflammatory Drugs
OCP	Organochlorine Pesticides
Oh	Ohmic Heating
PAHS	Polycyclic aromatic hydrocarbons
PEF	Pulsed Electric Field
PHE	Phenylalanine
PLE	Pressured Liquid Extraction
POH	Pulsed Ohmic Heating
Pro	Proline
PSE	Pressurized Solvent Extraction
PUFAs	Polyunsaturated Fatty Acids
RB	Rhodamine B
RDOC	Refractory Dissolved Organic Carbon
RF	Radiofrequency (RF)
ROS	Reactive Oxygen Species
RT	Reverse Transcriptase

SB	Sidi Bishir
SBM	Soybean Meal
SBSM	Seaweed-Based Sustainable Mariculture
SDGs	Sustainable Development Goals
Ser	Serine
SFE	Supercritical Fluid Extraction
SLE	Solid Liquid Extraction
SWE	Subcritical Water Extraction
T1DM	Type I diabetes mellitus
T2DM	Type II diabetes mellitus
TCA	Tricarboxylic Acid
Thr	Threonine
TMV	Tobacco Mosaic Virus
TNF-α	Tumor Necrosis Factor
TOC	Total Organic Carbon
TPS	El-Tabia pumping
TRP	Tryptophan
Tyr	Tyrosine
UAE	Ultrasound-Assisted Extraction
Val	Valine
VAP	Virus-Associated Protein
VOCs	Volatile Organic Compounds

List of Figures

List of Tables

Chapter 1
Classification, Distribution of Seaweeds, and Controlling Factors

1.1 Classification of Seaweeds

Seaweeds are multicellular photosynthetic eukaryotic organisms with different pigment compositions and that store food (Rhodophyta (red), Phaeophyceae or Ochrophyta (brown), and Chlorophyta (green) (Fig. 1.1) (Rashad and El-Chaghaby 2020; El Zokm et al. 2021; Ismail et al. 2025). The phycoerythrin and phycocyanin mask other pigments, and carotenoids, notably beta-carotene, lutein, and zeaxanthin. Chlorophyll *a* and *b* are responsible for pigmentation in Chlorophyta, and pigments contain chlorophylls a, c, and carotenoids, with fucoxanthin dominating, which are responsible for pigmentation in Phaeophyceae. Pigmentation in Phaeophyceae, to thick, leather-like seaweeds from 2–4 m long, to smaller species 30–60 cm long. Red seaweeds ranged from a few centimeters to about a metre in length; even though red seaweeds can occasionally be purple or even brownish red, botanists nevertheless classify them as Rhodophyceae due to additional traits. Similar in size to red seaweeds, green seaweeds are likewise tiny (Matos et al. 2024).

Seaweed lipid concentration ranged from 1 to 5% of dry mass. Seaweed is rich in proteins, essential amino acids, lipids, polysaccharides, dietary fibers, minerals (Ca, Fe, I, Mg, P, K, Zn, Cu, Mn, Se, and F), vitamins (A, B1, B2, B9, B12, C, D, E, and K), and polyphenols. Amino acids, such as aspartic acid, glutamic acid, and alanine in seaweed proteins, were strongly connected with the umami flavour of seaweeds, which evaluated the feasibility of extracting seaweed protein for taste enhancer purposes (Matos et al. 2024; Cadar et al. 2025). Seaweeds biosynthesize a variety of compounds with associated biological activities as part of their secondary metabolism, which is dedicated to the development of defensive and adaptive responses to environmental stresses (Ismail et al. 2025). Overall, aquatic algae have a low lipid content, ranging from 1 to 5% of their total dry weight. However, they contain a significant amount of polyunsaturated fatty acids (PUFAs). Other lipids, such as sterols, can also be found in a variety of heterogeneous molecules, including glycolipids and phospholipids (Belghit et al. 2017; Praveen et al. 2019). Seaweeds

G. M. El Zokm et al., *Rethinking Seaweeds via Applications and Technologies*,
Environmental Science and Engineering, https://doi.org/10.1007/978-3-032-14956-5_1

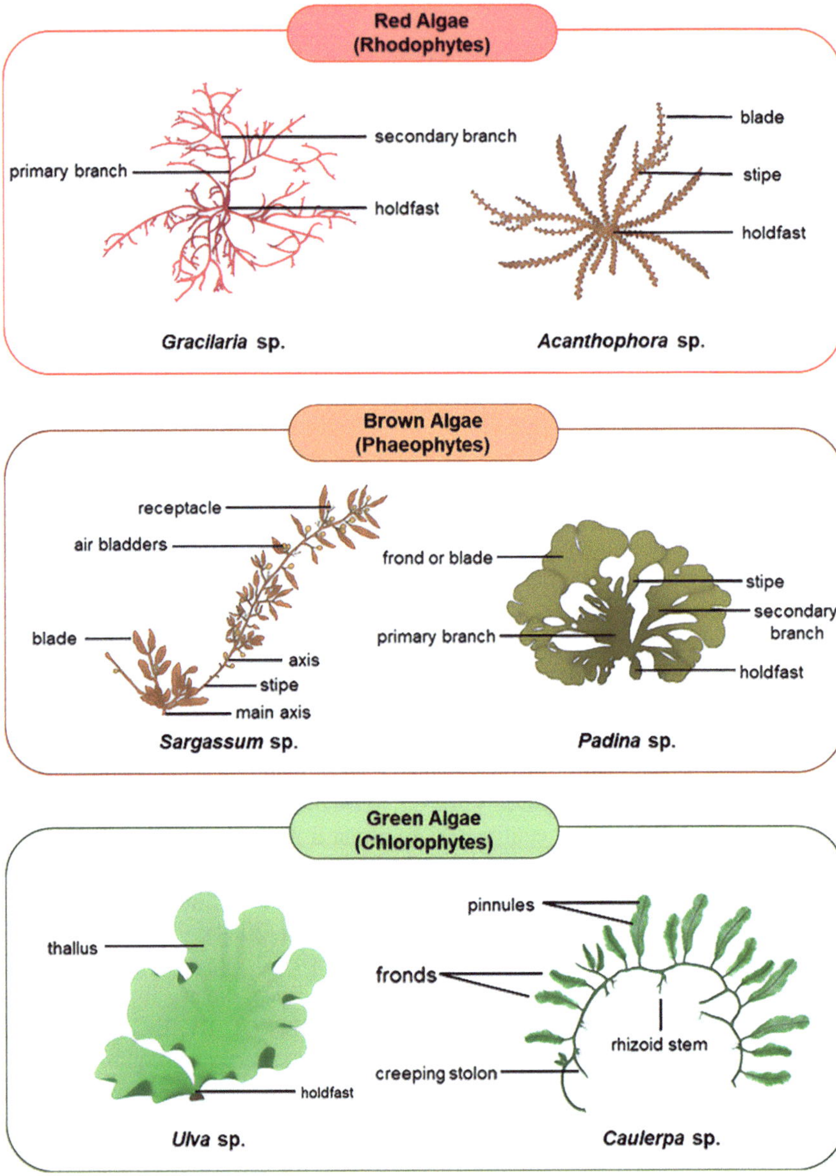

Fig. 1.1 Major groups of macroalgae and examples of their common species (Cheng et al. 2022)

provide a substantial dietary supply of protein and free amino acids (FAAs), as well as umami flavor, both of which are in great demand within the food industry. The predominant flavoring compounds used in the food industry are chemically manufactured, raising concerns about their safety and potential consumer resistance. These amino acids are frequently categorized based on their taste profiles: bitter (valine (Val), methionine (Met), isoleucine (Ile), leucine (Leu), phenylalanine (Phe), histidine (His), arginine (Arg), and tryptophan (Trp)), umami (aspartic acid (Asp) and glutamic acid (Glu)), tasteless (tyrosine (Tyr) and lysine (Lys)), and sweet (threonine (Thr), serine (Ser), glycine (Gly), alanine (Ala), and proline (Pro) (Ummat et al. 2021).

Seaweed region, harvesting time, and season all affect the amount and quality of physiologically active compounds found in seaweeds. Polyunsaturated fatty acids, polyphenols, carrageenan, and fucoidan are among the bioactive compounds that have been studied as novel sources from seaweeds and their extracts. Antioxidant, antimicrobial, antiviral, anticancer, antidiabetic, anti-inflammatory, anti-aging, anti-obesity, and antitumor properties have been demonstrated for these secondary metabolites. Moreover, seaweeds are the primary source of commercially important agar, carrageenan, and alginate (Mandal et al. 2025).

Recognizing that seaweed is the principal producer, it is at the bottom of the marine food web (Ismail and El Zokm 2025). Seaweeds and fellow humans have always been inextricably intertwined, primarily as a source of food, fibers, biochemicals, medicine, ornamental resources, artistic inspiration, and aesthetic values in a number of coastal populations. Furthermore, they are currently being investigated as green carbon, carbon sequestration, and a potential source for biological and pharmacological applications (Cotas et al. 2023). They are emerging as one of the alternative protein sources with several benefits over traditional high-protein crops in relation to nutritional value, productivity, protein yield per unit area, and the need for arable land, freshwater, and artificial fertilizer an important supplier of natural food additives, including colorants, flavoring agents, antioxidants or preservatives, and many more substances with health claims (Thiviya et al. 2022). Seaweeds form one of the important biotic components of different ecosystems. Examples of images of brown, green, and red seaweed species are given in Fig. 1.2. Many factors influence seaweed distribution, including physical (substrate, temperature, light quality and quantity, dynamic tidal activity, winds, and storms), chemical (salinity, pH, nutrients, gases, and pollution level), and biological (herbivores, microbes, epiphytes, endophytes, symbionts, parasites, and diseases).

All concerns regarding the formulated seafood analogs pertain to consumers' satisfaction with the raw food components, particularly in terms of nutritional, structural, and sensory properties, as well as the economic price of the end product. Even though the market for seafood analogs is growing as companies develop new products, algae and fungi appear to be increasingly relevant raw materials, not only as protein sources but also as sources of sensory qualities. As such, further investigation of algae and fungal sources is necessary, which are high in proteins and available in huge amounts (Karimidastjerd et al. 2025).

Phylum: Chlorophyta;
Class: Chlorophyceae;
Order: Ulvales;
Family: Ulvaceae
Genus (1) *Ulva lactuca* (Linnaeus)

Genus: (2) *Enteromorpha linza* (Linnaeus) Agardh

Phylum (2): Phaeophyta;
Class: Phaeophyceae;
Order: Dictyotales;
Family: Dictyotaceae;
Genus: (1) *Padina boryana* Thivy

Order (2): Fucales;
Family: Sargassaceae;
Genus: (1) *Sargassum wightii* Greville ex J.Agardh

Phylum: Rhodophyta, **Class:** Florideophyceae,
Order: (1) Corallinales,
Family: Corallinaceae,
Genus: (1) *Corallina mediterranea* Areschoug in J. Agardh

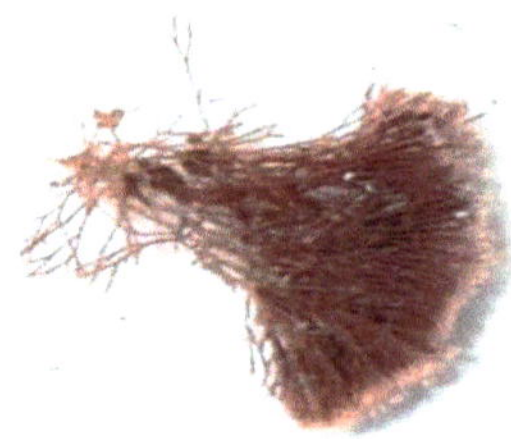

Genus: (2) *Jania rubens* (Linnaeus) Lamouroux

Fig. 1.2 Images of brown, green, and red seaweed species (Karimidastjerd et al. 2025)

As a result, a concise overview of the nutritional and sustainable features of algae is provided, based on (a) the sources' nutritional quality and health advantages; (b) maintaining the potential of algae and fungal sources; and (c) introducing market trends that are similar to those of seafood. Algae are essentially a useful material to use in seafood analogs due to their sensory qualities and health advantages.

Red seaweed has the highest protein content, comparable in quantitative terms to legumes at 30–40% dry matter, and its polysaccharides are sulfate galactans such as agar and carrageenan. Red seaweeds capture photons thanks to additional pigments such as phycoerythrin. So, it can grow deep even under low-light conditions. Brown contains only 15% protein; its primary polysaccharides are alginic acids, also known as alginates, which are salts of alginate. It captures photons due to the presence of additional pigments such as the carotenoid fucoxanthin. So, it can grow deep in low-light environments. It has only 30 percent protein. Green seaweeds have only 30% protein. Because of their additional pigments, such as chlorophyll, which is the primary light-absorbing pigment in green seaweed. Regardless, the differences caused by several aspects, including algae species, seasonality, and algae sections, must be acknowledged.

1.1.1 Red Seaweeds (Rhodophyta)

They are considered among the most ancient eukaryotic algae (macroalgae) that exist in both freshwater and marine environments. They have the most diverse taxonomic group, with roughly 6500 species found so far. Aside from food, red algae are mostly used as a source of gelling hydrocolloids, anticancer, anti-coagulant, antioxidant, and UV radiation uptake (Xie et al. 2023).

With a smaller morphology than green and brown algae, Rhodophyta are generally found in equatorial regions along intertidal zones and wider. Red macroalgae output reached over 9 million wet tons over the past ten years, accounting for 47% of the total yield of farmed macroalgae, according to the Food and Agriculture Organization (FAO) (Carpena et al. 2022) (Fig. 1.3).

Fig. 1.3 Photo of common red seaweed species with potential from Egyptian coasts (Ismail et al. 2020)

1.1.2 Phaeophyceae or Ochrophyta (Brown Seaweeds)

It is the biggest class of seaweeds, which are great suppliers of several beneficial chemicals. It has been utilized in traditional medicine for quite some time (Chen et al. 2024). This class includes 14 orders, 265 genera, and around 2000 species. Brown algae's crude polysaccharides have better antioxidant capabilities than those of red and green seaweeds, as they contain more polyphenols and sulfates (Nguyen and Tran 2024). Over the last two decades, the seaweed cultivation and production industry has experienced steady growth. According to Food and Agriculture Organization (FAO) figures, worldwide seaweed output has quadrupled from 118,000 to 358,200 tons. China leads the market, accounting for 58% of global aquaculture production (Zhang et al. 2022). It is well-known for its poor nutritional value, limited protein and fat levels, and greater mineral contents, while possessing around 1100 bioactive chemicals. It accounts for over 40% of worldwide macroalgae output, or about 10 million tons per year (Cadar et al. 2025).

Various types of them are utilized as food or added nutrients in people's daily meals. Long-term scientific investigations on patients have found links between an algae-rich diet and a lower risk of developing arterial hypertension, cardiovascular disease, cerebrovascular disease, and certain oncological disorders. A growing curiosity in algae as a rich source of bioactive metabolites, particularly, has led to promising applications in the manufacture of functional foods and nutraceuticals (Valado et al. 2023).

Phaeophyceae morphology: Kelps are dominating and crucial components of coastal ecosystems around the globe (Harley et al. 2012). Morphological studies have mostly focused on the frond width, narrow or broad, edge form, flat or ruffled, and overall frond surface shape, which are surface topographic characteristics in the central frond bullations or corrugations. However, the core front's surface form is complicated. When adjusting to diverse environmental conditions, there are several types of shapes and differential modifications throughout the frond. With the increasing global need for kelp as a source of food, feed, or energy. Other known and emergent uses of coastal regions are also increasing (Khan et al. 2024).

Fucales is one of the most diverse orders of the Phaeophyceae, containing the most common seaweeds in temperate and subtropical coastal regions from seven distinct families: *Sargassaceae, Fucaceae, Himanthaliaceae, Durvillaeaceae, Notheiaceae, Hormosiraceae,* and *Seirococcaceae.* Members are multicellular and feature the usual structure of big maritime macroalgae, including a well-differentiated rhizoid, stipe, and lamina. The blade is generally larger and may include gas vesicles to allow it to float freely in water. They are a large collection of seaweed species that vary greatly in terms of polyphenol (Fig. 1.4).

Phaeophyceae are a good source of essential minerals and vitamins, including retinol (A), B, ascorbic acid (C), and alpha-tocopherol (E) (Afonso et al. 2019). Brown algae can be used as a food reserve to fulfil the daily intake of some minerals like Sodium (Na), Potassium (K), Calcium (Ca) (Afonso et al. 2019). Magnesium (Mg) (Ferreira et al. 2022) and trace elements zinc (Zn), iron (Fe) (Skrzypczyk et al.

Fig. 1.4 Images of common brown seaweed species from Hurghada (Ismail et al. 2025)

2024), Manganese (Mn) (Ferreira et al. 2022), also have chromium (Cr) (Skrzypczyk et al. 2024), Iodine (I) (El Zokm et al. 2021), and Copper (Cu) (Ferreira et al. 2022).

1.1.3 Chlorophyta (Green Seaweeds)

They are the most prevalent macro seaweed species and represent a valuable marine biological resource. It has a high concentration of active chemicals that are utilized in antioxidant capacity, immunoregulation, and anti-inflammatory responses. They are often found in the intertidal zone. Common green seaweed species belong to *Ulva*, *Enteromorpha*, *Chaetomorpha*, *Codium*, and *Caulerpa* genera (Xu et al. 2022). Chlorophyta is usually eaten as a fresh vegetable (Syakilla et al. 2025). *Caulerpa lentillifera* (also known as "sea grapes" or "green caviar") is an edible green seaweed with a unique texture and a variety of nutritional advantages (Stuthmann et al. 2023) (Fig. 1.5).

Ulvophyceae, including *Ulva* (Ulvaceae, Ulvales), also known as sea lettuce, serve as a primary producer, supplying food and shelter for a variety of marine animals, as well as contributing to biofuel production, animal feed, and integrated aquaponics systems. However, its widespread blooms can harm tourism and the local economy. *Ulva* thallus is composed of either branching or unbranched monostromatic tubes, blade-like plants, blades with tubular borders, or all may coexist. Green algae blooms of the genus Ulva are occurring globally and are primarily attributed to anthropogenic factors (Mutizabal-Aros et al. 2024) (Figs. 1.6, 1.7, 1.8 and 1.9).

Fig. 1.5 Images of the selected macroalgal species on the Egyptian coast (El Zokm et al. 2021)

Fig. 1.6 Morphological variation of the thallus of Ulva stenophylloide (Mutizabal-Aros et al. 2024)

Green seaweeds are the most nutrient-dense species and excellent sources of micronutrients Potassium (K), Sodium (Na), and Phosphorus (P) (Xu et al. 2022). Copper (Cu), Zinc (Zn), and Cobalt (Co) were high in the green alga (Xu et al. 2022). Chlorophyta, **a** good source of magnesium (Mg), iron (Fe), and calcium (Ca) (Palaniyappan et al. 2023).

Chlorophyta are nutritionally rich in ascorbic acid (C) and hence commonly used for therapeutic purposes and as a dietary supplement. Also high in cobalamin (B12), which is required for brain function and many chemical processes in the human body (Xie et al. 2023). Thiamine (B_1), B_2 (Riboflavin), *Niacin* (B_3), and *Pyridoxine* (B_6), *Retinol* (A), and *Tocopherol* (E) (Cadar et al. 2025).

Fig. 1.7 Morphological variation of the thallus of *Ulva uncialis* (Mutizabal-Aros et al. 2024)

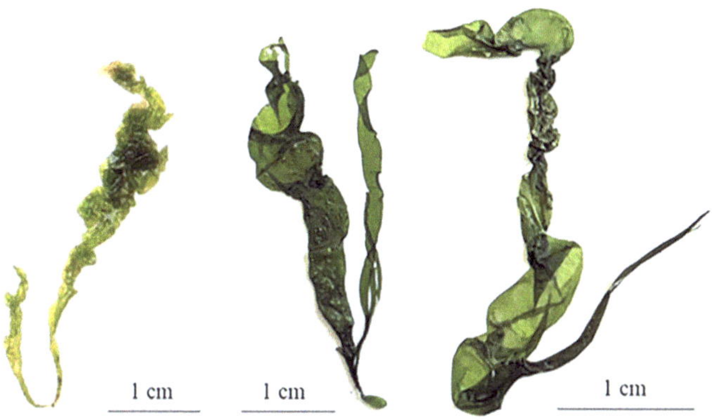

Fig. 1.8 Thallus of *Ulva compressa* (Mutizabal-Aros et al. 2024)

1.2 Distribution of Seaweeds and Controlling Factors

Seaweeds are found on beaches and firm substrates. Asian seaweed accounts for over 98% of total global production. China leads the globe in manufacturing, followed by Indonesia. Other countries that contribute to seaweed production include South Korea, the Philippines, Japan, and North Korea (Zhang et al. 2022). The distribution and abundance of seaweeds along the Egyptian coasts are sparse and scattered due to the large stretch of our coastline and the lack of focused research on this point.

Fig. 1.9 Thallus of *Ulva aragoensi* (Mutizabal-Aros et al. 2024)

Coll et al. (2010) recorded 187 accepted seaweed species from El-Arish to Salloum, including 107 red, 45 green, and 35 brown seaweeds. Recently, Farghaly (2018) determined 380 seaweed species from the Western Egyptian coastal area during the 2000–2010 period. The pattern of seaweeds' distribution is influenced by various factors, including physical parameters (substrate, temperature, light quality and quantity, dynamic tidal activity, winds, and storms), chemical parameters (salinity, nutrient levels, and pollution), and biological factors (epiphytes, microbes, endophytes, herbivores) (Fig. 1.10). Usually, the macroalgal vegetation and distribution were generally much more diverse in the period summer-autumn than in winter-spring (Madkour and El-Shoubaky 2007). Calcareous and thick leathery species were more distributed and abundant due to their structural complexity, with a relatively high degree of resistance to physical damage.

1.2.1 Physical Factors Affecting the Distribution of Macroalgae

1.2.1.1 Light Quality and Quantity

Figure 1.11 illustrated that Rhodophyta species are more abundant and have higher species richness in environments with low light availability compared to Chlorophyta species, which lose their competitive advantage in low-light environments. This may be due to their accessory pigment, like phycoerythrins in Rhodophyta and carotenoids, fucoxanthin in brown algae, which are responsible for capturing photons, but chlorophyll as the main light-absorbing pigment in Chlorophyta species (Branco et al. 2009; Dumay et al. 2014; Gómez and Huovinen 2020).

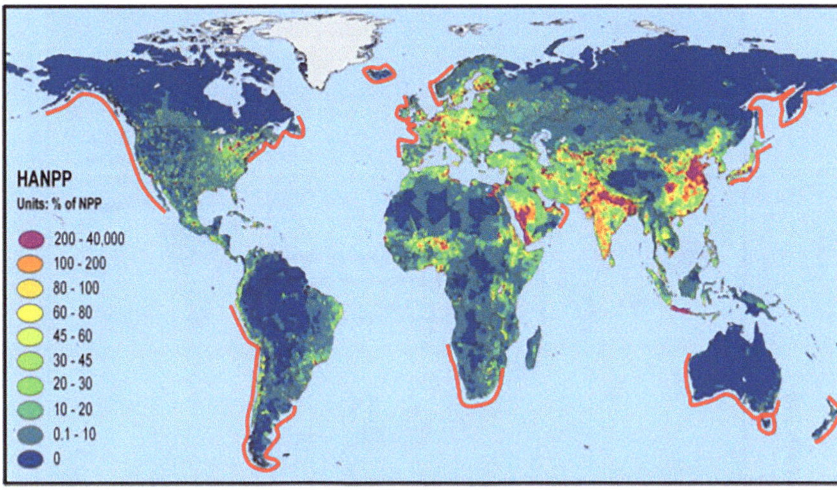

Fig. 1.10 Distribution of macroalgae (red line), signifying coastal regions with the capacity for macroalgae cultivation for biogas, alongside human appropriation of net primary output as a percentage of local net primary productivity (NPP) (Santelices 2007; Hughes et al. 2012)

1.2.1.2 Temperature

It affects the activity and distribution of the aquatic organisms (Fig. 1.12). Red seaweeds exhibited the strongest positive correlation with the gradient of maximum temperature (Sjøtun et al. 2015; Flores et al. 2023). Filamentous thalli and coarsely branching seaweeds exhibit higher productivity and thrive in temporally variable habitats, in contrast to thicker, calcareous seaweeds that are prominent in more stable conditions (Littler and Littler 1984). The optimal temperature at sea level and in saltwater for seaweed is 20–32 °C (Tulandi and Tumangkeng 2021).

1.2.1.3 Substrate

Many algal species are recorded as epiphytes that utilize other organisms as a substrate for physical support (Fig. 1.13) (Baweja et al. 2016; Nieder et al. 2019). In addition to being a direct food source for fish and sea urchins, seaweeds form the base of marine food webs. They also give fish and insects a place to live and a place to breed. The most common benthic creatures used as a gauge of the health of an ecosystem are seaweeds and coral animals. Environmental factors, both chemical and physical, have an impact on seaweeds. Seaweeds depend on the constant movement of water to absorb gases and nutrients from the surrounding environment. Seaweeds are also subjected to mechanical stress from the seawater's continuous circulation. Seaweeds have blades, a flexible stipe, and a robust holdfast to withstand

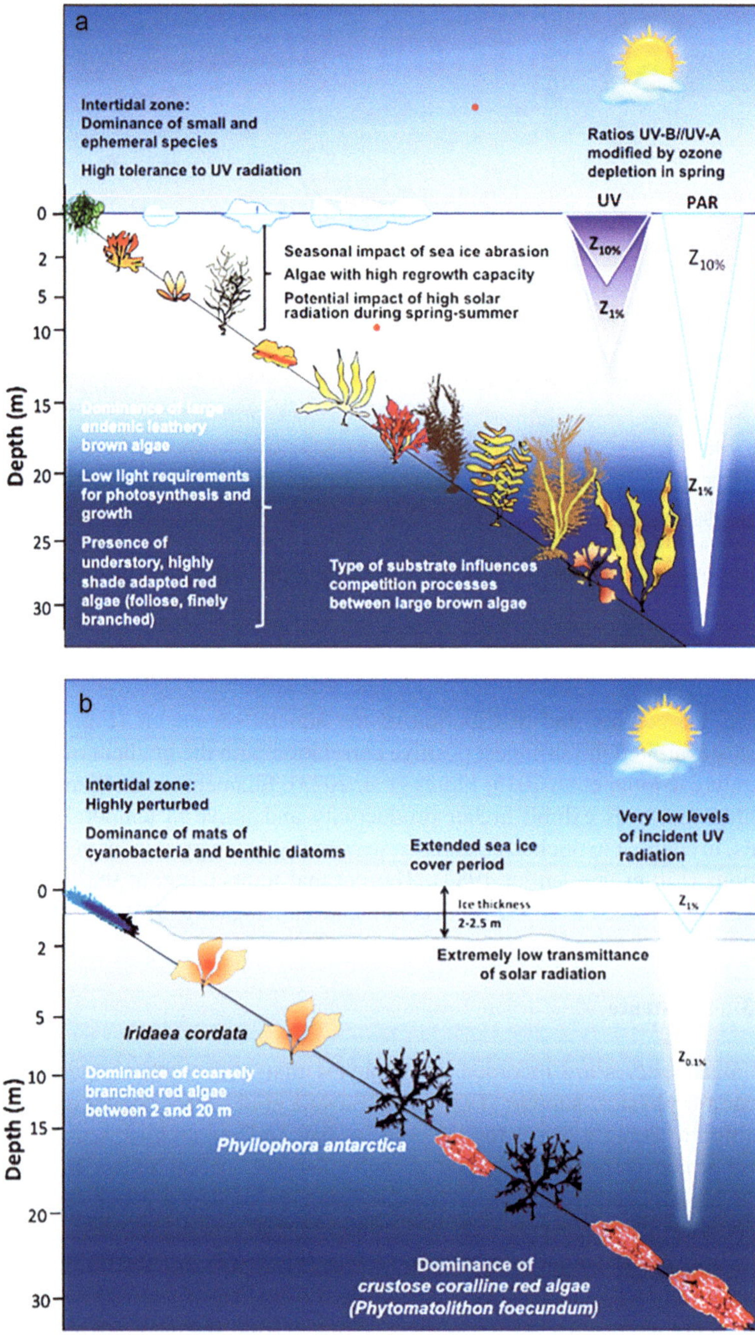

Fig. 1.11 Vertical distribution of macroalgae according to light (Gómez and Huovinen 2020)

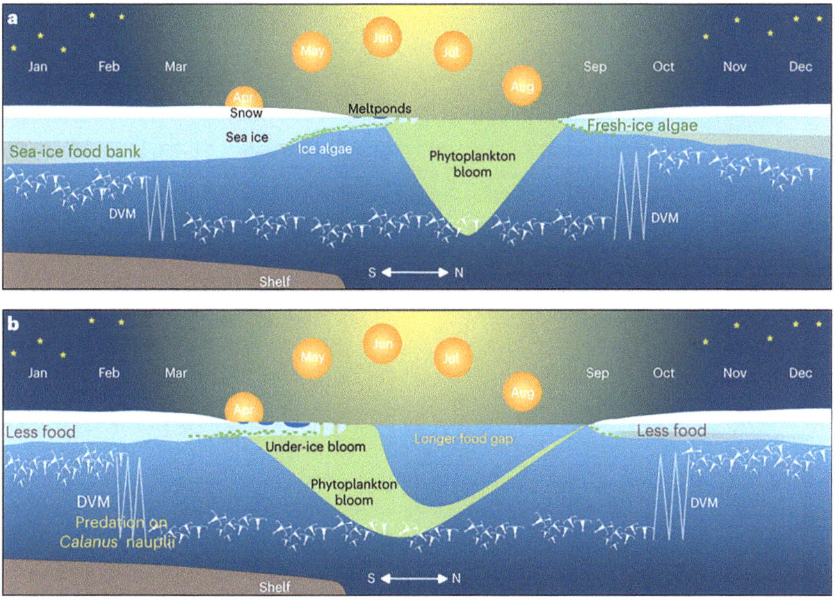

Fig. 1.12 Shifts in seasonal migratory timing and the commencement of DVM (diel vertical migration) can affect Phytoplankton in the Arctic Ocean's surface layer (0–50 m). The picture depicts an illustrated yearly cycle, with summer in the very middle (Flores et al. 2023)

this tension. Because of their elasticity, seaweeds can bend in the direction of the substratum when wave energy surrounds them.

Algal growth in the intertidal zone is shown to be most dependent on a stable substrate, and connection to a rough substrate promotes settling. Numerous man-made surfaces, such as logs, ropes, tires, bricks, and plastics, can be attached to seaweeds (Fig. 1.14). Many seaweeds, like *Laminaria digitata* and *Alaria esculenta*, exhibit additional morphological adaptations against strong water currents by growing robust blades due to a high number of cortical cells (Lee et al. 2017). Numerous seaweeds are frequently seen as epiphytes that grow on the surface of other species, including coral reefs, higher plants, animals, and other seaweeds (Fig. 1.14).

1.2.1.4 Epiphytic

Epiphytic Macroalgae on Other Macroalgae

Madkour and El-Shoubaky (2007) noticed that *Achrochaetum unifilum, Ceramium spp., Fosliella farinosa and Erythrotrichia carnea* (red seaweeds) grew on one or more algae without preference.

Macroalgae are found on rocks or other hard surfaces. The Red Sea's environment is diverse and abundant, due in part to the 2000 kms of coral reef that stretches from

Fig. 1.13 Seaweeds grow on a variety of substrates. **a** Natural rocky coast; **b** artificial rocky coast; **c** mangroves; **d** corals; **e** as an epiphyte on Rhodymenia; **f** and **g** on a rope; and **h** a car tire (Baweja et al. 2016)

the Gulf of Suez and Aqaba in the north to Bab El-Mandab and the Gulf of Aden in the south. It is home to a broad range of marine species, and its algae derive from the adjacent African and Arabian coastlines (Shams El-Din and Rashedy 2023).

Fig. 1.14 Epiphytic macroalgae that grow on G. divaricata. **a** Valonia ventricosa, **b** Caulerpa chemnitzia, **c** Dictyota sp., **d** Lobophora sp., and Hypnea caespitosa (Nieder et al. 2019)

Epiphytic Macroalgae on Mangrove

Mangrove ecosystems support diverse macroalgal communities that function as primary producers (Fig. 1.15). The community composition, macroalgae percentage cover, and physicochemical parameters of seawater were examined at two mangrove sites in the Red Sea, Egypt. A thick layer of macroalgae that is resilient to most environmental factors is maintained by mangrove trees. The health and purity of the water in the Red Sea are demonstrated by the stability of macroalgal assemblages in the two regions under investigation during the past 20 years (Rashedy et al. 2023).

Epiphytic Macroalgae on Coral Reefs

Macroalgae are linked to rocks or other hard substrates. The Red Sea has a diversified and rich ecology, with 2000 kms of coral reef that stretch from the Gulf of Suez and Aqaba in the north to the Bab El-Mandab and the Gulf of Aden in the south. It provides homes for a diverse variety of marine animals, and its algae (Fig. 1.16) come from the surrounding African and Arabian shores (Shams El-Din and Rashedy 2023).

Fig. 1.15 Macroalgae on a mangrove tree. **a, b** Rhizophora mangrove (Kalpitiya); **c, d** rhizophores of Rhizophora sp. on which epiphytic algae develop (Chilaw lagoon); **e** detail of a rhizophore covered by Caloglossa leprieurii and Polysiphonia sp. (Chilaw lagoon) (Coppejans et al. 2009)

1.2.1.5 Dynamic Tidal Activity

Tidal waste promotes the buildup of seaweed plants on shores, resulting in a eutrophication process and the spread of insects, nematodes, and bacteria that adversely affect the beach ecology and nearby human economic activity (Moral et al. 2023). The existence of flexible vegetation, such as seaweed, may diminish the hydrodynamic force of the entering bore by changing the flow characteristics (Oladokun 2023). Alterations in ocean circulation produce mild changes in species composition and activity, particularly among turf algae and upright seaweeds, rapid or abrupt, depending on the nature of the circulation changes (El-Manaway and Rashedy 2022; Paine et al. 2023a, b; Chen et al. 2024).

Fig. 1.16 (**a**) An algal bloom of Acanthophora spicifera on the Uva Island reef. (**b**) Cyanobacteria cover the algal thalli that protect the algae against herbivory (Fong and Paul 2011)

1.2.1.6 Sedimentation

The microscopic stage in the macroalgal life cycle is affected by a thin layer of silt since it can cut out the incident irradiance below a critical level (Norton 1978). With Wave action, they have a synergistic scouring effect on seaweed (Engledow and Bolton 1994).

1.2.1.7 Wave Action

Wave intensity led to 75% variation between algal communities of semi-exposed and sheltered sites (Blamey and Branch 2009). The nature of the macroalgae thallus and efficient attachment mechanism help it to persist in these stressful environments. *Gelidium* has a high attachment strength and erect fronds of fibrous consistency that allow it to resist strong beating from waves (Dudgeon and Johnson 1992).

Fluctuations in surface seawater carbon dioxide (CO_2) have a substantial impact on the overall metabolism of marine algae, particularly in the early phases of macroalgal growth (Paine et al. 2023a, b).

1.2.1.8 Sea Level Rise

The rate of sea level rise owing to thermal expansion of the seas and melting of glaciers and ice sheets is one to two millimeters per year. Within three algal groups, various species will have extremely varying colonization and dispersion potentials, resulting in very variable responses to the increase in accessible substrate with sea-level rise (Harley et al. 2006). The estimated pace of rise is sluggish in comparison to the life spans of most algal turfs, upright macroalgae, and CCA; therefore, strong rates of colonization, growth, and reproduction will lessen the susceptibility of macroalgae to sea-level rise (El-Manaway and Rashedy 2022).

1.2.1.9 Wind and Storms

The passive maritime dispersion of natural and farmed seaweeds is poorly understood, as are their floating and stranding dynamics. During the summer, floating kelps were promptly replenished from benthic sources to compensate for regular losses caused by deterioration and stranding. Because of the reduced losses, biomass remains high during the fall. Increased wind speed was an excellent predictor of biomass accumulations. At mid-latitudes, the dynamics of floating seaweed stocks depend on the availability of the benthic source population (Rothäusler et al. 2021).

Sargassum seaweed has expanded well beyond the Sargasso Sea, with large strandings on the shores of the West Indies and Mexico, posing major economic, ecological, and health risks. Atlantic pelagic algae can move in rafts. Their size and shape can change with the breeze. Using the OLCI/Sentinel-3 satellite, 1 million Sentinel-3 clumps were created across the West Indies. Analysis of the size distribution revealed that wind influences disaggregation and consolidation, depending on wind speed and aggregation size: Low winds promote agglomeration for the smallest aggregations and disaggregation for the biggest; High winds favor breakdown for all aggregation sizes (Laval et al. 2025).

Algal diel vertical migrations and transport obstacles allow algae to remain in the nearshore area. The storm-induced coastward spread of M. polykrikoides might raise the potential of future blooms in the coastal region (Xiong et al. 2022).

1.2.2 Chemical Factors Affecting the Distribution of Macroalgae

1.2.2.1 Salinity

Salinity is a limiting factor in macroalgal distribution (Fig. 1.17) (Sjøtun et al. 2015). Salinity and temperature are the major factors controlling the distribution of seaweeds in the Suez Canal (Farghaly and El-Shoubaky 2015). Red and brown seaweeds prevailed at relatively high salinities, while green seaweeds that contain many opportunists at low salinities (Sjøtun et al. 2015).

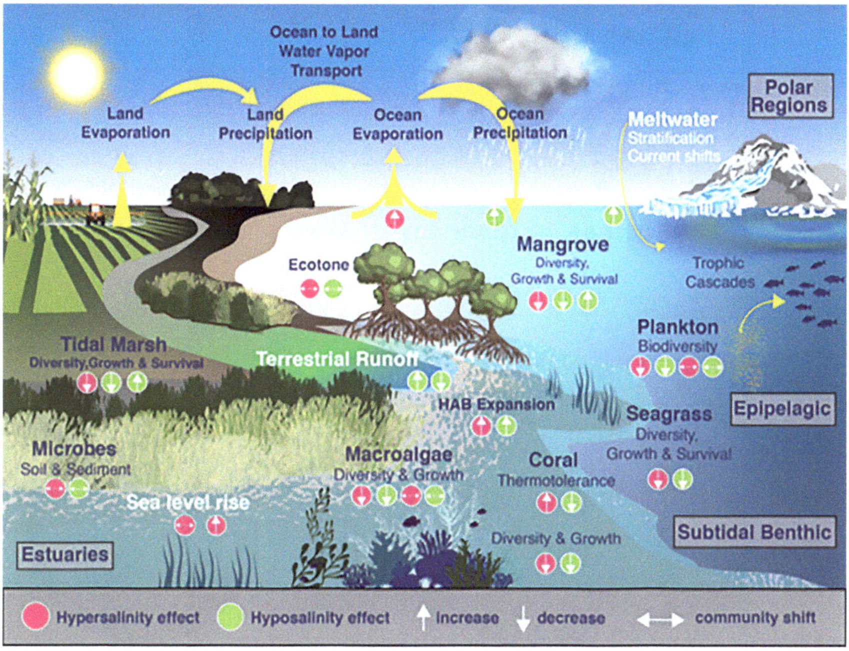

Fig. 1.17 Salinity fluctuations and their effects on ecosystem function (Röthig et al. 2023)

1.2.2.2 pH

The pH of saltwater has an impact on seaweed photosynthesis (Fig. 1.18). Generally, moderate pH values (about 6–8) are ideal for photosynthesis, with rates declining at both higher and lower pH levels. Carbon dioxide (CO_2) is a fundamental part of photosynthesis. In terrestrial environments, excess CO_2 can stimulate photosynthesis. However, in marine systems, where most macroalgae rely on bicarbonate ions (HCO_3^-) as a photosynthetic substrate, excessive concentrations in seawater can cause issues (Branch et al. 2013).

During Survival, development, and reproduction, all benthic macroalgae will be subject to fluctuations in pH, CO_2, and calcium carbonate saturation, which is especially relevant for crustose coralline algae (CCA), a form of red algae that deposits calcium carbonate and aids in reef consolidation. However, CCAs are especially vulnerable to ocean acidification. For example, reducing the pH from 8 to 7.5 significantly decreased calcification in the *alga Halimeda tuna* (El-Manaway and Rashedy 2022). Reducing the pH may also reduce the calcification of *Amphiroa* (Fig. 1.19) *and Corallina.* CCA is the algal group that is most susceptible to changes in carbonate saturation. Minor pH adjustments (from 8.1 to 7.8) lowered calcification by up to 21% for a coral reef ecosystem that includes CCA (Cornwall et al. 2017). A rise in pH caused an increase in dynamic viscosity, a decrease in antioxidant capacity, and lighter colored alginates (Nogueira et al. 2022).

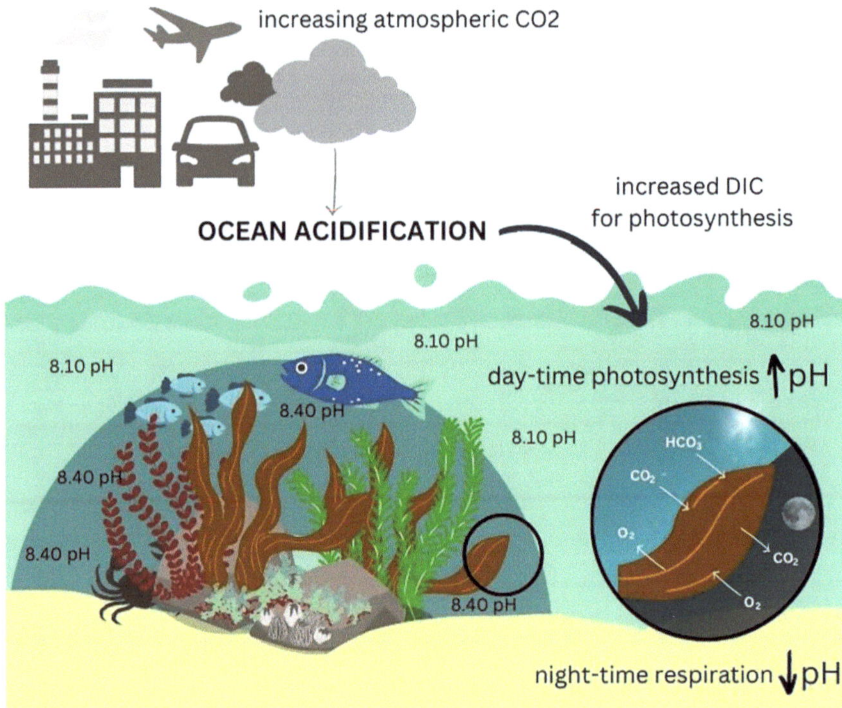

Fig. 1.18 The role of macroalgal habitats as ocean acidification refugia within coastal seascapes (Edworthy et al. 2023)

1.2.2.3 Nutrients

Nutrient concentrations have significant changes in macroalgal distribution (Fig. 1.20). Total phosphorus and nitrogen represented limiting growth and productivity factors of some marine macroalgae. Figueroa et al. (2009) noticed an increase in the macroalgal biomass in the Red Sea corresponding to nutrient enrichment from urban and aquaculture effluents and decreased herbivory. *Ulva* gives the maximum growth rate in enriched seawater by P and N (John 2013).

1.2.2.4 Pollution Level

Domestic and industrial pollutants (Fig. 1.21) might stimulate or inhibit the growth of algae (Shams El-Din et al. 2004; Nelson et al. 2008). El-Manawy and Shafik (2008) noticed that Chlorophyta growth increases vigorously in polluted habitats. Aleem (1993) listed more than 240 seaweed species from the Alexandria coast; that number has been greatly reduced in response to the impacts of pollution. Domestic

Fig. 1.19 *Amphiroa beauvoisii* of family Corallinaceae (Titlyanov et al. 2017)

wastes reduced the number of green and red algae in the Bitter Lakes (Suez Canal) (El-Manawy 1992).

The El-Tabia sediments contained the greatest quantities of Cd, Co, Ni, and Pb. Because of its proximity to the Abu Qir harbor, the Abu Qir area had the greatest quantities of the same heavy metals in the investigated water samples. This seaweed species accumulates trace elements; thus, they should be further examined for ecotoxicology if they are to be employed in the human diet (El-Mahrouk et al. 2023).

1.2.3 Biological Factors Affecting the Distribution of Macroalgae

1.2.3.1 Herbivores

The distribution and quantity of seaweeds may be influenced by connections among seaweed species, as well as between macroalgae and herbivores (Fig. 1.22). The effectiveness of marine herbivores as biological agents controlling invasive species is largely dependent on their feeding patterns, particularly the autotrophs they consume.

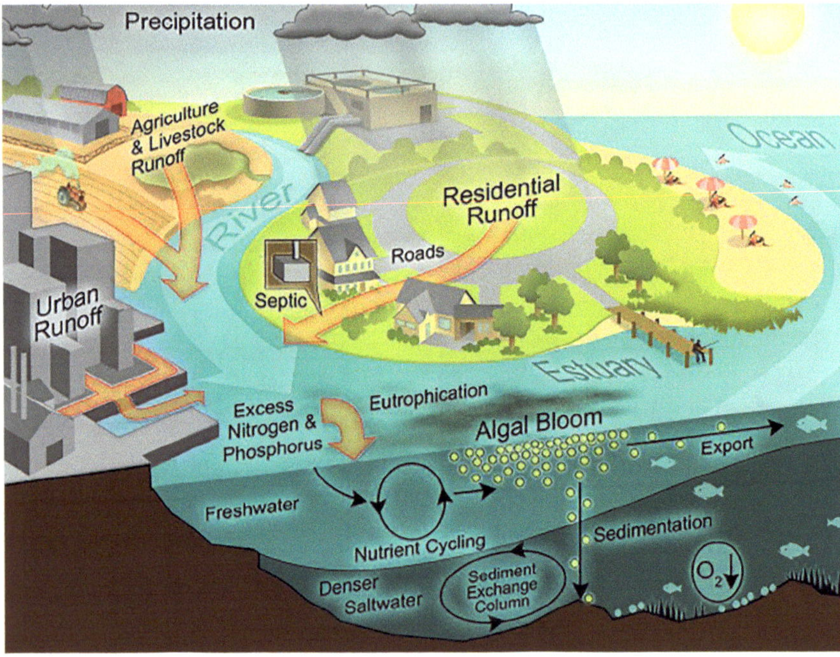

Fig. 1.20 Nutrient enrichment pathways that lead to eutrophication (Malone and Newton 2020)

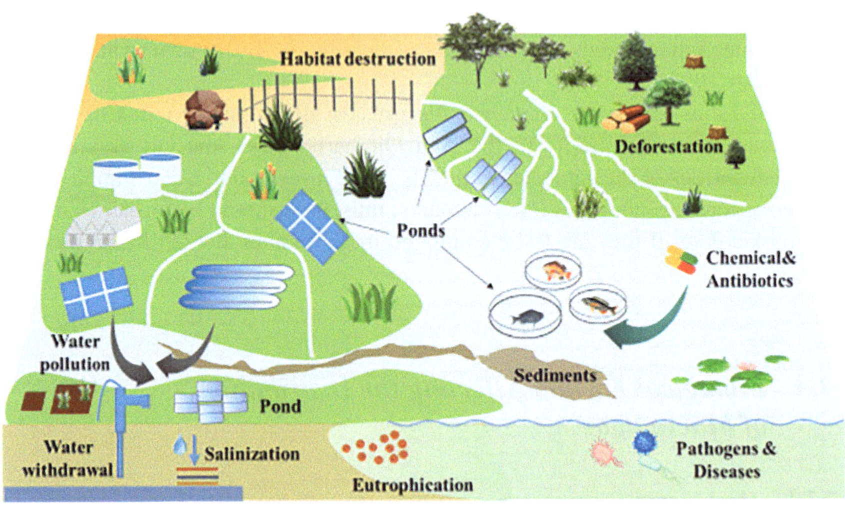

Fig. 1.21 Pollutants pathways (Liu et al. 2024)

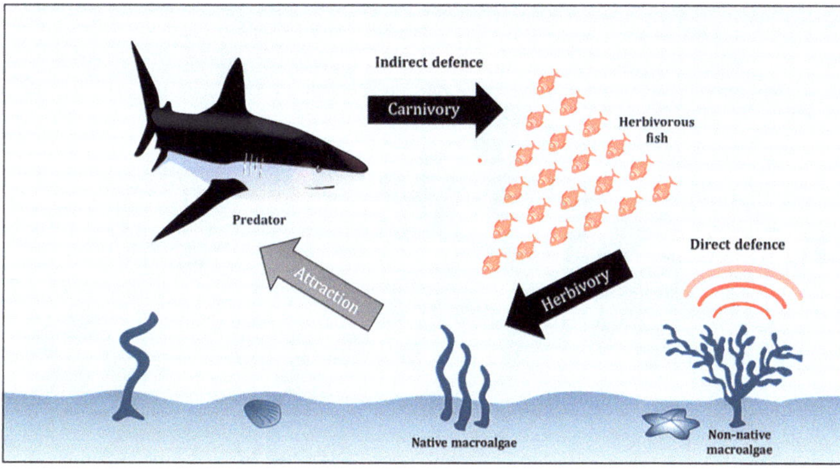

Fig. 1.22 Macroalgae-herbivore-predator interaction (Cheng et al. 2022)

Herbivore eating behavior can be changed by the nutritional characteristics, form, and novelty of autotrophs (Cheng et al. 2022).

Climate change is causing tropical and subtropical species to migrate to higher latitudes, which is connected to shoreline deforestation as a direct result of increased herbivorous fish, as well as the reduction of canopy-forming kelp. Utilization was focused on tropical and temperate Sargassum species, which are transitory and peak in abundance in the spring and early summer. Disturbance to algal competitors in shallow subtidal communities can promote the growth and dissemination of invasive and opportunistic local algae. If herbivores that depend on macroalgae typically shun it as a food source, then its continuous spread and rise in local abundance might trigger a fundamental change in ecological function to bottom-up management (Bradley et al. 2021). Management of Marine Protected Areas (MPAs), with international organizations intending to safeguard 30% of our seas by 2030. Herbivory incidence, rate, and frequency were examined in relation to protection status, depth, kelp species, and grazer type. In the MPA, herbivory is more prevalent, especially at deeper depths. Fish are the main kelp consumers, with a preference for S. polyschides, rather than urchins. Finding that fish are the main herbivores behind the growing loss in kelp biomass, with a predilection for *S. polyschides*, identifies fish as the major herbivores driving increasing kelp biomass decline (Arenas et al. 2024).

1.2.3.2 Symbionts (Holobiont)

Microbes

In the marine environment, seaweeds provide a wide range of ecological services and economic benefits (Saha et al. 2024). Microorganisms play critical roles in their health

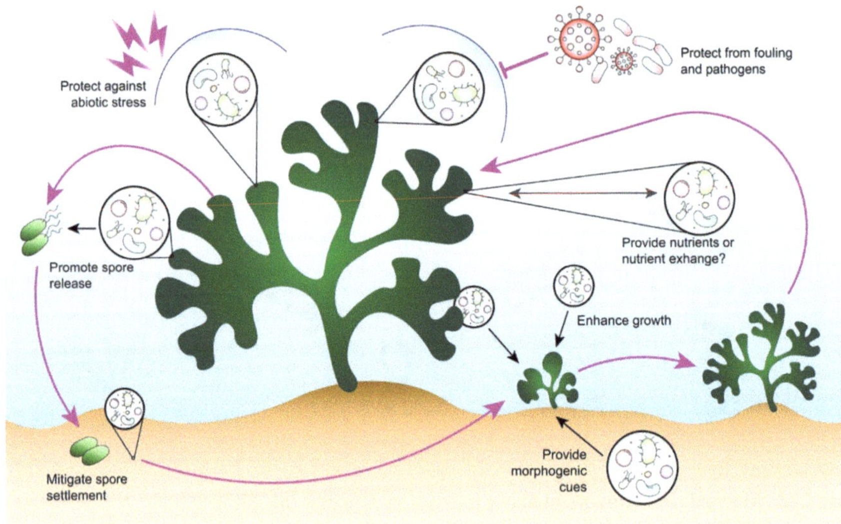

Fig. 1.23 Conceptual diagram showing multiple functions provided by microbial partners in the seaweed holobiont along its life cycle (represented by the purple arrows), from reproductive stage (spore release and settlement), morphogenesis and growth to physiological adaptive responses upon biotic (protection from pathogens and fouling) and abiotic stress. The black arrows indicate nutrient, metabolite, or cue fluxes between microbial partners and the algal host (Saha et al. 2024)

and functioning, suggesting the need to examine seaweeds and their microbiome as a cohesive unit or "holobiont" (Marzinelli et al. 2024) (Fig. 1.23). Various maritime seaweed species have been proven to contain unique bacterial populations (Syukur et al. 2024).

Epiphytic bacteria can play a significant part in controlling seaweed features, although their exact contributions to host development and natural products are unknown for most seaweeds. The epiphytic bacteria in A. taxiformis cultures can be modified to improve seaweed biomass output, although there may be a trade-off in natural product levels in some culture lines (Blanco et al. 2025).

In the Egyptian Red Sea, the single ascidian Phallusia nigra supported a symbiotic amphipod and four copepod species, with concentrations of up to 68 mixed symbionts per host. Correlation analysis revealed no rivalry or antagonism among symbionts (Sherif & Lombardi 2022).

The surface microbiome of seaweed hosts is a multi-domain biofilm that is controlled by host-microbe and microbe-microbe interactions. Host influence and promiscuity may play a key part in seaweed hosts acclimating to changing environmental circumstances and effectively flourishing in new environments (Bonthond et al. 2021).

Invasive species coexist with microbiota from their native range and interact with microbiota present in the new habitat to which they are transferred. Host adaptability toward microbiota, often known as host promiscuity, is a crucial feature. If exposed to the same circumstances, different populations form more similar or related microbial communities (Bonthond et al. 2021).

UV light has the potential to alter the physiology of seaweeds and the microorganisms that live on them. UV radiation altered the makeup of microbial communities on thalli, resulting in a decrease in community diversity and evenness. During UV exposure, the relative abundance of *Alteromonas sp. rose,* whereas that of *Methylophaga sp.* and *Colwellia sp.* declined (Dobretsov et al. 2020).

1.2.3.3 Epiphytic

Epiphytic Microorganisms on Macroalgae

The bacteria that settle on the seaweeds' surfaces and produce a wide variety of inorganic and organic compounds regulate the seaweeds' growth and development in this specific environment. To determine the epiphytic and endophytic bacterial populations linked to 10 common marine macroalgae from different locations along Alexandria's Mediterranean Sea coast. While edophytic bacteria linked to red algae declined in the Eastern Harbor site and grew at the Abu Qir and Miami locations, those linked to green algae rose at all sites (Abdelrazek et al. 2024). Additionally, to keep track of species and identify any potential changes to the ecosystem's physicochemical and biological properties, routine biodiversity monitoring is required (Sami et al. 2025).

Epiphytic Diatoms

The most prevalent kind of microalgae connected to the submerged hydrophyte *Myriophyllum spicatum* L. was diatoms. 66–95% (Abdel-Aal 2021).

Epiphytic Invertebrates

Epiphytic species greatly enhance ecological diversity and nutrient cycles by looking at the quantity and composition of the invertebrates associated with seaweeds in Egypt's Red Sea. In three different locations—Kalawy Bay, the Mangrove region, and the National Institute of Oceanography and Fisheries (NIOF) in Hurghada—six seaweed species—Caulerpa racemosa, Sargassum cinereum, Cystoseira myrica, Halimeda tuna, Hormophysa cuneiformis, and Digenia simplex—were collected. At three locations, 26 species of invertebrates were found. While nematodes were the most prevalent species linked to C. myrica, H. cuneiformis, and S. cinereum at Kalawy Bay, amphipods were the most prevalent fauna related to C. myrica and H. cuneiformis at the Mangrove area station. Amphipods were the most prevalent on C. myrica, while nematode species were the most pervasive at the NIOF station. C. myrica had the greatest species richness value at the Mangrove station, whereas

Sargassum cinereum had the highest values in the Kalaw Bay and NIOF stations, respectively. *D. simplex* and S. cinereum exhibited the most resemblance based on the abundances of linked epiphytic fauna (Omer et al. 2021).

1.2.3.4 Endophytes

Endophytic Fungi or Bacteria

Marine endophytic fungi have been found in algae, both intercellular and intracellular. By coevolution, specific interaction systems emerged because they influence the life cycle and homeostasis of their eukaryotic hosts. Organic signals from their host stimulate gene expression in endophytes, resulting in the generation of endophytic secondary metabolites that serve several biological activities for ecological sustainability (El-Bondkly et al. 2021). Sponges were collected at diverse depths from several locations. The isolated endophytes from these sponges show a wide range of microbial communities. Fungal filtrates effectively generate Copper Nanoparticles that exhibit moderate antibacterial activity against *Staphylococcus aureus* and *Escherichia coli*, with potential as antifungal agents in medical and healthcare settings. Morphological and microscopic investigations of Aspergillus sp. aid in identifying and characterizing the fungal species (Hassan et al. 2024).

1.3 Case Studies for the Distribution of Seaweeds (Green-Red-Brown) in Egypt

1.3.1 Case Study 1

In April 2011, El-Said and El-Sikaily (2013) collected a variety of seaweed species from the Mediterranean shore of Alexandria, Egypt. El-Said and El-Sikaily (2013) reported the collection of 18 algal samples representing three algal groups over 3 years (2008–2010) along the Egyptian Mediterranean coast (Alexandria). The seaweeds were collected from seven locations (Abu Qir Bay, El Montazah, Sidi Bishir, El Shatby, Eastern Harbor, El Mex Bay, and 21 km) and belonged to several classes, including red (*Jania rubens, Gracilaria compressa, Gracilaria verrucosa, Pterocladia capillacea, and Hypnea musciformis*), green (*Ulva lactuca, Codium tomentosum, and Enteromorpha intestinalis*), and brown (*Colpomenia sinuosa and Sargassum linifolium*).

1.3.2 Case Study 2

Mofeed and Deyab (2015) gathered macroalgal groups from five locations along the Suez Canal (Port Said, Qantara, Ismailia, Fayed, and Suez) over the course of a year. 34 macroalgal species were reported by the authors (14 Chlorophyta, 12 Phaeophyta, and 8 Rhodophyta). The three intermediate locations (Qantara, Ismailia, and Faied) were inhabited by Phaeophyta species, while Suez and Port Said were dominated by Chlorophyta. In contrast, Chlorophyta outnumbered Phaeophyta and Rhodophyta in terms of macroalgal plant abundance at three locations (Suez, Fayed, and Port Said), with Suez containing 89% of the total macroalgal vegetation, Fayed 51%, and Port Said 44%. Meanwhile, the other macroalgal community in Ismailia was filled by *Rhodophyta.*

1.3.3 Case Study 3

El Zokm et al. (2021), investigated concentration, pathways, and interactions of halogens relative to 22 key variables in fresh marine algae; Thirteen seaweed species from various divisions were collected from Alexandria, Egypt's Eastern Harbor. These included *Cladophora prolifera* (Roth) Kützing, *Cl. glomerata* (Linnaeus) Kützing, *Enteromorpha prolifera, Padina pavonica* (Müller) J. Agardh, *Ulva fasciata* Delile, *U. linza Linnaeus* and *U. lactuca* Linnaeus from Chlorophyceae, *Amphiroa fragilissima* (Linnaeus) J.V. Lamouroux, *Corallina officialis* Linnaeus, *Gigartinia teedi* (Mertens ex Roth) J.V. Lamouroux, and *Jania rubens* (Linnaeus) Lamouroux from Rhodophyceae. Consequently, this study can serve as a foundation for additional monitoring studies and a deeper comprehension of the paths of several elements in algae.

1.3.4 Case Study 4

Mediterranean sites—Eastern Harbor (EH), Abu Talat (AT), El Mex Bay (EM), Sidi Bishir (SB), and El-Tabia pumping (TPS)—were investigated in June 2019. Five Mediterranean sites—Eastern Harbor (EH), Abu Talat (AT), El Mex Bay (EM), Sidi Bishir (SB), and El-Tabia pumping (TPS)—were investigated in June 2019 in order to clarify the diversity trends in macroalgae populations. Five different types of macroalgae were discovered along the shore of Alexandria: *Ulva Fasciata, Ulva compressa, Corallina officinalis, Carollina elongate, and Petrocladia capillacea.* Alpha and beta diversity patterns were assessed at various heavy metal deposition levels and physicochemical factors in macroalgal environments. We discovered notable regional variations in alpha diversity and a modest beta index value (average ~ 0.6), which can be attributed to the wide range of heavy metal pollution. Compared to other macroalgae, Ulva fasciata is more abundant. The site with the highest diversity index value (H') is EH. The richness index of species (D), the highest evenness index (E), and the Simpson index are 1.28, 0.33, 0.59, and 0.69, respectively (AbouGabal et al. 2022).

1.3.5 Case Study 5

The research sought to assess the spatial variations in the biochemical composition of seaweed species along the coast of Alexandria, concentrating on fatty acids and total amino acids of the dominant edible species. Sixteen distinct species from the coast of Alexandria, Egypt, were examined, and the fatty acid and total amino acid compositions of the three predominant dietary species were evaluated. *Ulva fasciata* exhibits the largest moisture content (78.0% ± 9.9% fresh weight) in El Mex Bay, whereas *Jania rubens* possesses the biggest ash content (28.2% ± 2.9% dry weight) in Gleem, and *Pterocladia capillacea* demonstrates a significant carbohydrate content (52.5% ± 5.2% dry weight). A significant protein content was noted in U. linza (Abu-Qir). The lipid content in most seaweeds is minimal, and the caloric value is approximately 5 kcal per 100 g, rendering it a feasible alternative biosource of nutritious food to aid in combating obesity. The high photosynthetic pigment content of *Cladophora glomerata* and *U. linza* sets them apart.

The maximum amount of fucoxanthin is found in *Padina boryana.* Only *P. capillacea* has the largest amount of PUFA (5.26% ± 0.74%), suggesting that it has the best nutritional indices. High essential amino acid (EAA)/NEAA ratios were observed in *P. boryana* and *P. capillacea* (0.65 ± 0.35 and 0.63 ± 0.69, respectively). Lysine has a low amino acid score (AAS) of 62.20% ± 0.44%, making it the limiting amino acid in *P. boryana.* EAAI values for *P. capillacea* and Padina boryana were 122% ± 0.86% and 154% ± 0.95%, respectively. Fatty acids, total amino acids, and other biochemical compositions of the seaweed under investigation may find application in biofuel, medicine, cosmetics, and nutrition.

References

Abdel-Aal EI (2021) Species composition and diversity of epiphytic microalgae on *Myriophyllum spicatum* in the El-Ibrahimia Canal, Egypt. Afr J Aquat Sci 46(3):319–328. https://doi.org/10.2989/16085914.2021.1884523

Abdelrazek HM, Shams El-Din NG, Ghozlan HA et al (2024) Distribution and functional perspective analysis of epiphytic and endophytic bacterial communities associated with marine seaweeds, Alexandria shores, Egypt. BMC Microbiol 24:293. https://doi.org/10.1186/s12866-024-03426-x

AbouGabal AA, Khaled AA, Aboul-Ela HM, Aly HM, Diab MH, Shalaby OK (2022) Marine macroalgal biodiversity, spatial study for the Egyptian. Mediterranean Sea, Alexandria Coast

Afonso NC, Catarino MD, Silva AMS, Cardoso SM (2019) Brown macroalgae as valuable food ingredients. Antioxidants 8(9):365. https://doi.org/10.3390/antiox8090365

Aleem AA (1993) The Marine algae of Alexandria. Alexandria, Egypt., p 139

Arenas BF, Marques AF, Franco JN (2024) Potential backfiring effects of Marine protected areas on kelp herbivory. Sci Rep 14(1):1–12. https://doi.org/10.1038/s41598-024-82557-7

Baweja P, Kumar S, Sahoo D, Levine I (2016) Biology of seaweeds. In: Fleurence J, Levine I (eds) Seaweed in health and disease prevention. Academic Press, pp 41–106. https://doi.org/10.1016/B978-0-12-802772-1.00003-3

Belghit I, Rasinger JD, Heesch S et al (2017) In-depth metabolic profiling of marine macroalgae confirms strong biochemical differences between brown, red, and green algae. Algal Res 26:240–249. https://doi.org/10.1016/j.algal.2017.08.001

Blamey LK, Branch GM (2009) Habitat diversity relative to wave action of rocky shores: implications for the selection of marine protected areas. Aquat Conserv Mar Fresh Ecosyst 19(6):645–647

Blanco S, Campbell AH, Cummins SF et al (2025) Manipulation of the surface-associated bacteria on *Asparagopsis taxiformis* and its effect on seaweed growth and halogenated natural products. J Appl Phycol. https://doi.org/10.1007/s10811-025-03530-z

Bonthond G, Bayer TAS, Stärck N, Wang G, Nakaoka M, Künzel S, Weinberger F (2021) The role of host promiscuity in the invasion process of a seaweed holobiont. ISME J 15(6):1668–1679. https://doi.org/10.1038/s41396-020-00878-7

Bradley DJ, Boada J, Gladstone W, Glasby TM, Gribben PE (2021) Sublethal effects of a rapidly spreading native alga on a key herbivore. Ecol Evol 11(18):12605–12616. https://doi.org/10.1002/ece3.8005

Branch TA, DeJoseph BM, Ray LJ, Wagner CA (2013) Impacts of ocean acidification on marine seafood. Trends Ecol Evol 28:78–186

Branco CCZ, Krupek RA, Peres CK (2009) Ecological distribution of stream macroalgal communities from mid-western region of Paraná State. Braz Arch Biol Technol 52:379–386

Cadar E, Popescu A, Dragan AM, Pesterau AM, Pascale C, Anuta V, Prasacu I, Velescu BS, Tomescu CL, Bogdan-Andreescu CF, Sirbu R, Ionescu AM (2025) Bioactive compounds of marine algae and their potential health and nutraceutical applications: a review. Mar Drugs 23(4):152. https://doi.org/10.3390/md23040152

Carpena M, Garcia-Perez P, Garcia-Oliveira P, Chamorro F, Otero P, Lourenço-Lopes C, Cao H, Simal-Gandara J, Prieto MA (2022) Biological properties and potential of compounds extracted from red seaweeds. Phytochem Rev 22(6):1–32. https://doi.org/10.1007/s11101-022-09826-z

Chen D, Li G, Tang J, Zhao L, Wang YP, Gao C (2024) Evaluating the effects of laver cultivation on tidal flat erosion: toward sustainable environmental practices. J Environ Manage 366:121830. https://doi.org/10.1016/j.jenvman.2024.121830

Cheng A, Lim WY, Lim P, Yang Amri A, Poong S, Song S, Ilham Z (2022) Marine autotroph-herbivore synergies: unravelling the roles of macroalgae in marine ecosystem dynamics. Biology 11(8):1209. https://doi.org/10.3390/biology11081209

Coll M, Piroddi C, Steenbeek J, Kaschner K, Ben Rais Lasram F, Aguzzi J et al (2010) The biodiversity of the mediterranean sea: estimates, patterns, and threats. PLoS ONE 5(8):1–36

Coppejans E, Leliaert F, Dargent O, Gunasekara R, Clerck O (2009) Sri Lankan seaweeds: methodologies and field guide to the dominant species

Cornwall CE, Comeau S, McCulloch MT (2017) Coralline algae elevate pH at the site of calcification under ocean acidification. Glob Change Biol 23:4245–4256. https://doi.org/10.1111/gcb.13673

Cotas J, Gomes L, Pacheco D, Pereira L (2023) Ecosystem services provided by seaweeds. Hydrobiology 2(1):75–96. https://doi.org/10.3390/hydrobiology2010006

Dobretsov S, Véliz K, Romero MS, Tala F, Thiel M (2020) Impact of UV radiation on the red seaweed *Gelidium lingulatum* and its associated bacteria. Eur J Phycol 56(2):129–141. https://doi.org/10.1080/09670262.2020.1775309

Dudgeon SR, Johnson AS (1992) Thick vs. thin: thallus morphology and tissue mechanics influence differential drag and dislodgement of two co-dominant seaweeds. J Exp Mar Biol Ecol 165:23–43

Dumay J, Morançais M, Munier M, Le Guillard C, Fleurence J (2014) Chapter eleven—phycoerythrins: valuable proteinic pigments in red seaweeds. Adv Bot Res 71:321–343
Edworthy C, Steyn P-P, James NC (2023) The role of macroalgal habitats as ocean acidification refugia within coastal seascapes. Cambridge Prisms: Coastal Futures 1:e22. https://doi.org/10.1017/cft.2023.9
El Zokm GM, Ismail MM, El-Said GF (2021) Halogen content relative to the chemical and biochemical composition of fifteen marine macro and micro algae: nutritional value, energy supply, antioxidant potency, and health risk assessment. Environ Sci Pollut Res 28:14893–14908. https://doi.org/10.1007/s11356-020-11596-0
El-Bondkly EAM, El-Bondkly AAM, El-Bondkly AAM (2021) Marine endophytic fungal metabolites: a whole new world of pharmaceutical therapy exploration. Heliyon 7(3):e06362. https://doi.org/10.1016/j.heliyon.2021.e06362
El-Mahrouk ME, Dewir YH, Hafez YM, El-Banna A, Moghanm FS, El-Ramady H, Mahmood Q, Elbehiry F, Brevik EC (2023) Assessment of bioaccumulation of heavy metals and their ecological risk in sea lettuce (Ulva spp.) along the Coast Alexandria, Egypt: implications for sustainable management. Sustainability 15(5):4404. https://doi.org/10.3390/su15054404
El-Manawy IM, Shafik MA (2008) Morphological characterization of Halimeda (Lamouroux) from different biotopes on the Red Sea coral reefs of Egypt. Am-Eurasian 3:251–263
El-Manaway IM, Rashedy SH (2022) The ecology and physiology of seaweeds: an overview. In: Ranga Rao A, Ravishankar GA (eds) Sustainable global resources of seaweeds, vol 1. Springer, Cham. https://doi.org/10.1007/978-3-030-91955-9_1
El-Manawy IM (1992) Ecological studies on the marine benthic flora of the Bitter Lakes (Suez Canal). Ph.D. thesis, Suez Canal University, p 155
El-Said GF, El-Sikaily A (2013) Chemical composition of some seaweeds from the Mediterranean Sea coast, Egypt. Environ Monit Assess 185:6089–6099. https://doi.org/10.1007/s10661-012-3009-y
Engledow HR, Bolton JJ (1994) Seaweed a-diversity within the lower eulittoral zone in Namibia-the effects of wave action, sand inundation, mussels and limpets. Bot Mar 37:267–276
Farghaly MS (2018) Diversity of seaweeds and seagrasses in the Egyptian Mediterranean coastal eco-zones. J Coast Zone Manag 21:23
Farghaly MS, El-Shoubaky GA (2015) Synopsis of biodiversity and distrebution of macrophytes along the Suez Canal in time and space. In: International conference on plant, marine and environmental sciences (PMES-2015). Kuala Lumpur, Malaysia
Ferreira M, Salgado JM, Fernandes H, Peres H, Belo I (2022) Potential of red, green and brown seaweeds as substrates for solid state fermentation to increase their nutritional value and to produce enzymes. Foods 11(23):3864. https://doi.org/10.3390/foods11233864
Figueroa FL, Martínez B, Israel A, Neori A, Malta E, Ang P Jr, Inken S, Marquardt R, Rachamim T, Arazi U, Frenk S, Korbee N (2009) Acclimation of Red Sea macroalgae to solar radiation: photosynthesis and thallus absorptance. Aquat Biol 7:159–172. https://doi.org/10.3354/ab00186
Flores H, Veyssière G, Castellani G, Wilkinson J, Hoppmann M, Karcher M, Valcic L, Cornils A, Geoffroy M, Nicolaus M, Niehoff B, Priou P, Schmidt K, Stroeve J (2023) Sea-ice decline could keep zooplankton deeper for longer. Nat Clim Chang 13:1122–1130. https://doi.org/10.1038/s41558-023-01779-1
Fong P, Paul VJ (2011) Coral reef algae. In: Dubinsky Z, Stambler N (eds) Coral reefs: an ecosystem in transition. Springer, Dordrecht. https://doi.org/10.1007/978-94-007-0114-4_17
Gómez I, Huovinen P (2020) Form and function in antarctic seaweeds: photobiological adaptations, zonation patterns, and ecosystem feedbacks. In: Gómez I, Huovinen P (eds) Antarctic seaweeds. Springer, Cham. https://doi.org/10.1007/978-3-030-39448-6_11
Harley CDG, Anderson KM, Demes KW, Jorve JP, Kordas RL, Coyle TA, Graham MH (2012) Effects of climate change on global seaweed communities. J Phycol 48(5):1064–1078
Harley CDG, Hughes AR, Hultgren KM, Miner BG, Sorte CJB et al (2006) The impacts of climate change in coastal marine systems. Ecol Lett 9:228–241

Hassan MG, Elshafie A, Shahin FSA, Baraka DMM, Hamed AA (2024) Exploring the biotechnological potential of marine sponges: insights into endophytic microbial diversity, fungal biosynthesis of copper nanoparticles, and antimicrobial properties. Benha J Appl Sci 9(3):161–168. https://doi.org/10.21608/bjas.2024.278618.1368

Hughes AD, Kelly MS, Black KD, Stanley MS (2012) Biogas from Macroalgae: is it time to revisit the idea? Biotechnol Biofuels 5(1):86. https://doi.org/10.1186/1754-6834-5-86.PMID: 23186536;PMCID:PMC3542030

Ismail MM, Lopez JMM, El-Sayed AAM (2025) Evaluation of the spatial variations in the biochemical composition of seaweed species along the coast of Alexandria, with a focus on fatty acids and total amino acids of the prevalent edible species. Food Sci Nutr 13:e70645. https://doi.org/10.1002/fsn3.70645

Ismail MM, Alotaibi BS, El-Sheekh MM (2020) Therapeutic uses of red macroalgae. Molecules 25(19):4411. https://doi.org/10.3390/molecules25194411

Ismail MM, Zokm GME (2025) Algae as keystone for blue economy: sustainability and challenges. Discov Sustain 6:366. https://doi.org/10.1007/s43621-024-00746-w

John B (2013) Microalgae for biofuels and animal feeds. Energies 6:5869–5886

Karimidastjerd A et al (2025) Nutritional and sustainability aspects of algae and fungi sources in seafood analogs—a review. Ann Anim Sci 25(2):417–428. https://doi.org/10.2478/aoas-2024-0072

Khan N, Sudhakar K, Mamat R (2024) Macroalgae farming for a sustainable future: navigating opportunities and driving innovation. Heliyon 10(7):e28208. https://doi.org/10.1016/j.heliyon.2024.e28208

Laval M, Aimene Y, Descloitres J, Courtrai L, Zongo P, Dorville R, Chevalier C (2025) The influence of wind on the spatial distribution of pelagic sargassum aggregations in the tropical Atlantic. Water 17(6):776. https://doi.org/10.3390/w17060776

Lee WK, Lim YY, Leow ATC, Namasivayam P, Abdullah JO, Ho CL (2017) Factors affecting yield and gelling properties of agar. J Appl Phycol 29:1527–1540

Littler MM, Littler DS (1984) Relationships between macroalgal functional form groups and substrata stability in a subtropical rocky intertidal system. J Exp Mar Biol Ecol 74:13–34

Liu X, Wang Y, Liu H, Zhang Y, Zhou Q, Wen X, Guo W, Zhang Z (2024) A systematic review on aquaculture wastewater: pollutants, impacts, and treatment technology. Environ Res 262:119793. https://doi.org/10.1016/j.envres.2024.119793

Madkour F, El-Shoubaky GA (2007) Seasonal distribution and community structure of macroalgae along port said coast, Mediterranean Sea, Egypt. Egypt J Aquat Biol Fish 11(l):221–236

Malone TC, Newton A (2020) The globalization of cultural eutrophication in the coastal ocean: causes and consequences. Front Mar Sci 7:670

Mandal AK, Parida S, Behera AK, Adhikary SP, Lukatkin AA, Lukatkin AS, Jena M (2025) Seaweed in the diet as a source of bioactive metabolites and a potential natural immunity booster: a comprehensive review. Pharmaceuticals 18(3):367. https://doi.org/10.3390/ph18030367

Marzinelli EM, Thomas T, Gonzalez SV, Egan S, Steinberg PD (2024) Seaweeds as holobionts: current state, challenges, and potential applications. J Phycol 60(4):785–796. https://doi.org/10.1111/jpy.13485

Matos J, Cardoso C, Serralheiro ML, Bandarra NM, Afonso C (2024) Seaweed bioactives potential as nutraceuticals and functional ingredients: a review. J Food Compos Anal 133:106453. https://doi.org/10.1016/j.jfca.2024.106453

Mofeed J, Deyab M (2015) Monitoring for the abundance and distribution of macroalgae along Suez Canal, Egypt. 11(1):81–91

Moral A, Greyer V, Aguado R, Tijero A (2023) Material revalorization of beach wrack comprising seaweed and marine phanerogams: optimization of hydrothermal treatments. Biomass Bioenerg 174:106856. https://doi.org/10.1016/j.biombioe.2023.106856

Mutizabal-Aros J, Ramírez ME, Haye PA, Meynard A, Pinilla-Rojas B, Núñez A, Latorre-Padilla N, Search FV, Tapia FJ, Saldías GS et al (2024) Morphological and molecular identification of Ulva

spp. (ulvophyceae; chlorophyta) from Algarrobo Bay, Chile: understanding the composition of green tides. Plants 13:1258. https://doi.org/10.3390/plants13091258
Nelson TA, Haberlin K, Nelson AV, Ribarich H et al (2008) Ecological and physiological cont rols of species composition in green macroalgal blooms. Ecology 89(5):1287–1298
Nguyen NTT, Tran TM (2024) A review: classification, chemical compositions and antioxidant properties of red, brown and green macroalgae. ILMU KELAUTAN Indones J Mar Sci 29(3):340–350. https://doi.org/10.14710/ik.ijms.29.3.340-350
Nieder C, Liao C-P, Chen C, Liu S-L (2019) Filamentous calcareous alga provides substrate for coral-competitive macroalgae in the degraded lagoon of Dongsha Atoll, Taiwan. PLoS ONE 14:e0200864. https://doi.org/10.1371/journal.pone.0200864
Nogueira MT, Chica LR, Yamashita C, Nunes NSS, Moraes ICF, Branco CCZ, Branco IG (2022) Optimal conditions for alkaline treatment of alginate extraction from the brown seaweed Sargassum cymosum C. Agardh by response surface methodology. Appl Food Res 2(2):100141. https://doi.org/10.1016/j.afres.2022.100141
Norton TA (1978) The factors influencing the distribution of *Saccorhiza polyschides* in the region of Longh Ine. J Mar Biol Assoc UK 58:527–536
Oladokun OS (2023) Hydrodynamic model tests for seaweed as a source of energy reduction during extreme events. Aquac J 3:181–195. https://doi.org/10.3390/aquacj3030015
Omer MF, Abd El-Wakeil KNM, Hussein H, Rashedy SH (2021) Invertebrate assemblages associated with seaweeds at different locations in the Red Sea, Egypt. Egypt J Aquat Biol Fish 25(1):407–421. https://doi.org/10.21608/ejabf.2021.143933
Paine ER, Britton D, Schmid M, Brewer EA, Diaz-Pulido G, Boyd PhW, Hurd CL (2023a) No effect of ocean acidification on growth, photosynthesis, or dissolved organic carbon release by three temperate seaweeds with different dissolved inorganic carbon uptake strategies. ICES J Mar Sci 80(2):272–281. https://doi.org/10.1093/icesjms/fsac221
Paine ER, Boyd PW, Strzepek RF et al (2023b) Iron limitation of kelp growth may prevent ocean afforestation. Commun Biol 6:607. https://doi.org/10.1038/s42003-023-04962-4
Palaniyappan S, Sridhar A, Kari ZA, Téllez-Isaías G, Ramasamy T (2023) Evaluation of phytochemical screening, pigment content, in vitro antioxidant, antibacterial potential and GC-MS metabolite profiling of green seaweed *Caulerpa racemosa*. Mar Drugs 21(5):278. https://doi.org/10.3390/md21050278
Praveen MA, Parvathy KRK, Balasubramanian P, Jayabalan R (2019) An overview of extraction and purification techniques of seaweed dietary fibers for immunomodulation on gut microbiota. Trends Food Sci Technol 92:46–64. https://doi.org/10.1016/j.tifs.2019.08.011
Rashad S, El-Chaghaby GA (2020) Marine Algae in Egypt: distribution, phytochemical composition and biological uses as bioactive resources (a review). Egypt J Aquat Biol Fish 24(5):147–160. https://doi.org/10.21608/ejabf.2020.103630
Rashedy SH, Hamed ESAE, El-Manawy IM, Pereira L (2023) Stability of macroalgal assemblage in the mangrove swamp as an indicator for the health and quality of the Red Sea waters. Egyptian J Aquat Res. https://doi.org/10.1016/j.ejar.2023.01.004
Rothäusler E, Hinojosa IA, Moraga J, Pizarro-Koch M, Ramos M, Thiel M (2021) At the mercy of the winds: the seasonal dynamics of floating and stranded seaweeds at mid-latitudes. Limnol Oceanogr 66(12):4391–4402. https://doi.org/10.1002/lno.11969
Röthig T, Trevathan-Tackett SM, Voolstra CR, Ross C, Chaffron S, Durack PJ, Warmuth LM, Sweet M (2023) Human-induced salinity changes impact marine organisms and ecosystems. Glob Change Biol 29(17):4731–4749. https://doi.org/10.1111/gcb.16859
Saha M, Dittami SM, Chan CX, Raina B, Stock W, Ghaderiardakani F, Valathuparambil Baby John AM, Corr S, Schleyer G, Todd J, Cardini U, Bengtsson MM, Prado S, Skillings D, Sonnenschein EC, Engelen AH, Wang G, Wichard T, Brodie J et al (2024) Progress and future directions for seaweed holobiont research. New Phytol 244(2):364–376. https://doi.org/10.1111/nph.20018
Sami M, Ahmed F, Temraz TA et al (2025) Ecological study on seaweed diversity in Suez, Hurghada, and Marsa Alam, Red Sea, Egypt. BMC Ecol Evo 25:52. https://doi.org/10.1186/s12862-025-02389-5

Santelices B (2007) The discovery of kelp forests in deep-water habitats of tropical regions. Proc Natl Acad Sci 104:19163–19164

Shams El-Din NGT, Rashedy SH (2023) Recent introduced algal species in the Egyptian Marine waters. In: Biodiversity of seaweeds in the Egyptian Marine waters. Earth and environmental sciences library. Springer, Cham. https://doi.org/10.1007/978-3-031-33366-8_4

Shams El-Din NG, El-Moselhy KhM, Amer A (2004) Distribution of some macroalgae in the Intertidal zone of the Suez bay in relation to environmental conditions. Egypt J Aquat Res 30:171–188

Sherif M, Lombardi T (2022) Spatial variability in a symbiont-diverse marine host and the use of observational data to assess ecological interactions. Diversity 14(3):197. https://doi.org/10.3390/d14030197

Sjøtun K, Husa V, Asplin L, Sandvik AD (2015) Climatic and environmental factors influencing the occurrence and distribution of macroalgae—a fjord gradient revisited. Mar Ecol Prog Ser 532:1–39

Skrzypczyk VM, Callahan DL, Francis DS et al (2024) Australian brown seaweeds as a source of essential dietary minerals. J Appl Phycol 36:797–809. https://doi.org/10.1007/s10811-023-03074-0

Stuthmann LE, Brix da Costa B, Springer K et al (2023) Sea grapes (Caulerpa lentillifera J. Agardh, Chlorophyta) for human use: structured review on recent research in cultivation, nutritional value, and post-harvest management. J Appl Phycol 35:2957–2983. https://doi.org/10.1007/s10811-023-03031-x

Syakilla N, George R, Matanjun P (2025) The potential of noodles incorporated with green seaweed Caulerpa lentillifera as a low-glycaemic food. J Appl Phycol 37:2965–2974. https://doi.org/10.1007/s10811-025-03564-3

Syukur S, Richmond J, Majzoub ME, Nappi J, Egan S, Thomas T (2024) Not all parents are the same: diverse strategies of symbiont transmission in seaweeds. Environ Microbiol 26(1):e16564. https://doi.org/10.1111/1462-2920.16564

Thiviya P, Gamage A, Gama-Arachchige NS, Merah O, Madhujith T (2022) Seaweeds as a source of functional proteins. Phycology 2(2):216–243. https://doi.org/10.3390/phycology2020012

Titlyanov E, Titlyanova T, Li X, Huang H (2017) Common marine algae of Hainan Island (Guidebook), pp 75–228. https://doi.org/10.1016/B978-0-12-811963-1.00004-4

Tulandi DA, Tumangkeng JV (2021) Analysis of daily dynamics of thermal interaction of temperature and ocean flow in seaweed growth areas. J Phys Conf Ser 1968:012035

Ummat V, Garcia-Vaquero M, Poojary MM et al (2021) Green extraction of proteins, umami and other free amino acids from brown macroalgae *Ascophyllum nodosum* and *Fucus vesiculosus*. J Appl Phycol 33:4083–4091. https://doi.org/10.1007/s10811-021-02569-y

Valado A, Ponedilok P, Caseiro A, Pereira L (2023) Macroalgae as a source of functional foods in the prevention of cardiovascular diseases. Recent Prog Nutr 3(4):024. https://doi.org/10.21926/rpn.2304024

Xie C, Lee ZJ, Ye S, Barrow CJ, Dunshea FR, Suleria HAR (2023) A review on seaweeds and seaweed-derived polysaccharides: nutrition, chemistry, bioactivities, and applications. Food Rev Intl 40(5):1312–1347. https://doi.org/10.1080/87559129.2023.2212055

Xiong J, Shen J, Wang Q (2022) Storm-induced coastward expansion of Margalefidinium polykrikoides bloom in Chesapeake Bay. Mar Pollut Bull 184:114187. https://doi.org/10.1016/j.marpolbul.2022.114187

Xu N, Wang W, Xu K, Xu Y, Ji D, Chen C, Xie C (2022) Cultivation of different seaweed species and seasonal changes cause divergence of the microbial community in coastal seawaters. Front Microbiol 13:988743. https://doi.org/10.3389/fmicb.2022.988743

Zhang L, Liao W, Huang Y, Wen Y, Chu Y, Zhao C (2022) Global seaweed farming and processing in the past 20 years. Food Prod Process Nutr 4:23

Chapter 2
Primary and Functional Secondary Metabolites in Seaweed with Special Reference to "Omics" Technologies

2.1 Seaweed (Macroalgae) Metabolites

Metabolites are the final products of metabolic processes facilitated by numerous enzymes present in cells. They are characterized as biochemical intermediates; their concentrations might be considered the definitive response of biological systems to genetic or environmental alterations. Physiological processes like reproduction are directly impacted by primary metabolites under typical growth settings. On the other hand, secondary metabolites are endogenous substances generated in reaction to a variety of stressors, such as exposure to ultraviolet [UV] light, temperature and salinity fluctuations, or environmental contaminants (Rosa et al. 2020). The term "metabolome" refers to the entire collection of metabolites, which includes both endogenous and exogenous metabolites. While the body produces endogenous metabolites, it does not normally produce exogenous metabolites, which include poisons, dietary additives, and chemical pollutants. The two categories of endogenous metabolites are primary metabolites (like amino acids, lipids, and carbohydrates) that are involved in growth, development, and reproduction, and secondary metabolites (like flavonoids, phenolics, terpenoids, and alkaloids) that are not directly related to growth (Fig. 2.1).

A subfield of "omics," metabolomics is the thorough qualitative and quantitative examination of every metabolite in a biological system at a particular moment in time (Sánchez-Machado et al. 2004; Kumar et al. 2017; Padam 2020; Premarathna et al. 2022; Carpena et al. 2023). Metabolomics can be used to investigate how storage conditions, geographic location, and even an organism's reaction to various environmental stressors can all affect bioactivity (Xu et al. 2024; Varadharajan et al. 2025). The present state of seaweed metabolomics, or biochemical composition, was discussed in this chapter along with the primary and bioactive secondary metabolite composition of red, brown, and green algae. The primary metabolites of red and brown algae were alanine, threonine, citrate, 3PG, glutamic acid, valine, and

G. M. El Zokm et al., *Rethinking Seaweeds via Applications and Technologies*, Environmental Science and Engineering, https://doi.org/10.1007/978-3-032-14956-5_2

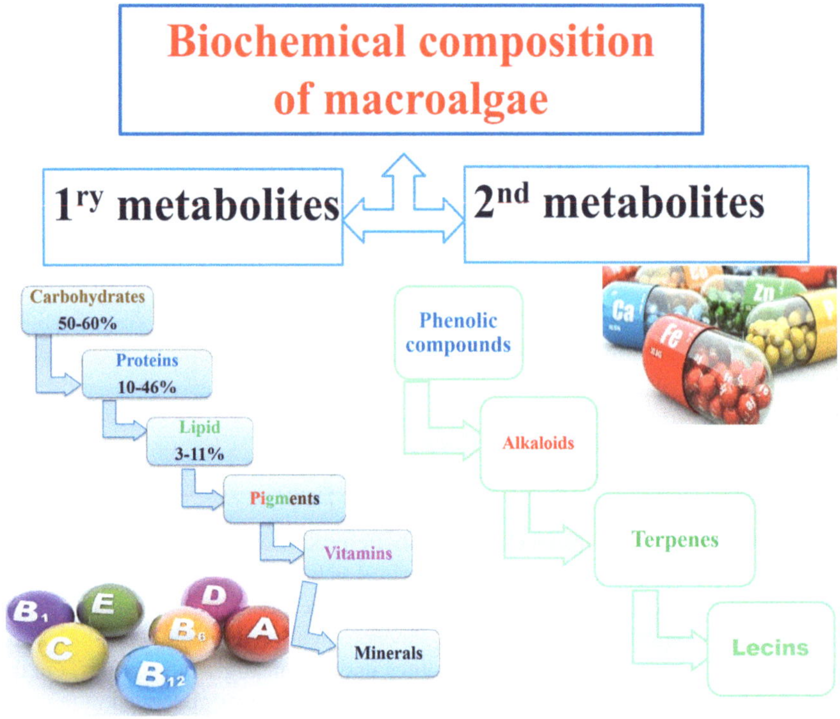

Fig. 2.1 Seaweed metabolomics (Own elaboration)

isoleucine, while the primary metabolites of green algae were galactose and mannose (Park et al. 2023).

Comprehensive metabolic profiling, also known as metabolomics, offers comprehensive details about an organism's biochemical phenotype. Metabolomics has many uses, including chemotaxonomy, environmental monitoring, and the identification of natural products, in addition to enhancing our comprehension of biochemical processes. Seaweeds are among the most significant primary producers in the aquatic environment; however, there aren't many comparative large-scale metabolomics investigations for marine macroalgae (Belghit et al. 2017).

Seaweeds are morphologically and functionally diverse creatures that are typically categorized into three classes based on their pigmentation: green algae (Chlorophyta), red algae (Rhodophyta), and brown algae (Phaeophyceae, Heterokontophyta). Over the past five decades, seaweeds have become a significant source of active metabolites utilized in pharmaceutics, nutrition, bioenergy, agriculture, and cosmetics, among other sectors. The intricacy of these creatures is demonstrated by the more than 3000 chemicals that have been identified in seaweeds. Two key elements are believed to be responsible for this: (a) special evolutionary traits and (b) harsh environmental conditions. The endosymbiont theory states that a prokaryotic photosynthetic cyanobacterium and a non-photosynthetic eukaryotic protist host underwent an initial endosymbiosis to produce the red and green algae (Cavalier-Smith 2000; Chan and Bhattacharya 2010). On the other hand, because brown algae evolved via a secondary endosymbiosis event combining a non-photosynthetic eukaryote and a unicellular red alga, they are a phylogenetically distant lineage from the red and green algae.

Metabolites are biochemical precursors or end products of spontaneous metabolic events catalyzed by various enzymes within cells, and their levels can be considered as biological systems' ultimate response to genetic or environmental changes. The metabolome is the total set of metabolites that includes both endogenous and exogenous compounds. Endogenous metabolites are classified as main metabolites, which are responsible for growth, development, and reproduction, such as amino acids, lipids, and carbohydrates, and secondary metabolites, which are not directly related to growth. Exogenous metabolites [such as chemical pollutants, food additives, and poisons]. They are not created naturally by the body (Xu et al. 2024; Varadharajan et al. 2025).

2.1.1 Primary Metabolites

These basic metabolites are divided into four categories: carbohydrates, lipids, proteins, and minerals, which are necessary for seaweed survival and contribute to its nutritional and medicinal value. The major metabolite content of each of these types in seaweed varies greatly depending on the species, geographical location, and season (Marinho-Soriano et al. 2006; Khairy and El-Shafay 2013; Balboa et al. 2016). The main primary metabolite classes in green, brown, and red seaweeds are:

Carbohydrates make up to 40% of seaweed biomass. The protein concentration of seaweeds varies depending on their species, growing conditions, season, and geographic location. Protein concentrations in brown, green, and red seaweeds vary from 4–10%, 15–25%, and 8–40%, respectively. Because they contain important fatty acids like omega-3, lipids are of tremendous interest. Seaweed has a lower lipid content than other terrestrial plants. Brown seaweed usually has a lipid content of 1–4.5%.

2.1.1.1 Carbohydrates

While seaweeds, like Fucus spiralis, may contain free simple carbohydrates with high levels of the alcoholic sugar mannitol, they are primarily polysaccharides, primarily non-starch and non-homopolymers that can be linear or branched, and involve monosaccharides (primarily aldohexoses and aldopentoses) where the terminal CH_2OH group can be replaced by methyl or carboxylic groups. Monosaccharides bonded by a glycosidic linkage also have different possibilities [α or β, $1 \rightarrow 3$, $1 \rightarrow 4$, $1 \rightarrow 6$], … Seaweed polysaccharides are structurally very distinct and characteristic of each genus of algae due to the random combination of these possibilities.

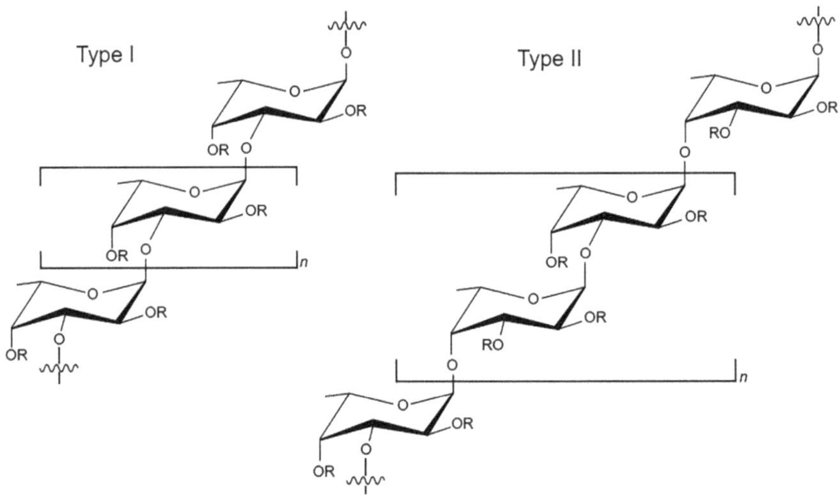

Fig. 2.2 Schematic representation of fucoidan type I and type II structure. R can be α-Lfucopyranose, α-D-glucuronic acid, and sulfate groups; the location of galactose, mannose, xylose, rhamnose, arabinose, and glucose depends on each seaweed species or the circle of life

Sulfated (fucoindans, carragenans, and ulvan) and non-sulfated (alginates and agars) polysaccharides are the most well-known and used in industry. Some, such as laminarin, xylans, porphyrans, argassan, and floridean, are less common, have no industrial applications at present, and are still being studied (Seca et al. 2018; Tanna and Mishra 2018; Xie et al. 2023).

Seaweed-type Polysaccharides (Islam et al. 2022; Akter et al. 2024).

Green: Ulvan, starch, xyloglucan, cellulose, hemicellulose, xylopyranose.

Brown: Fucoidan, laminaran, alginates, cellulose.

Red: Agar, cellulose, carrageenan, agaropectin, xylans, mannans.

Fucoidans

Brown seaweeds contain a family of L-fucose polysaccharides called fucoidans, which have varying percentages of sulfate groups and various glycosidic connections. Fucoidans perform a structural role akin to that of terrestrial plant hemicelluloses. The molecular weight, location, and amount of the sulfate groups, as well as the existence of additional minor monosaccharides in the fucoidan structure, determine their advantageous effects. Since fucoidans seem to lessen the obvious signs of aging skin, such as wrinkles, freckles, and blemishes, their usage in cosmetic products has already been documented in the literature (Fig. 2.2) (Usov et al. 2022).

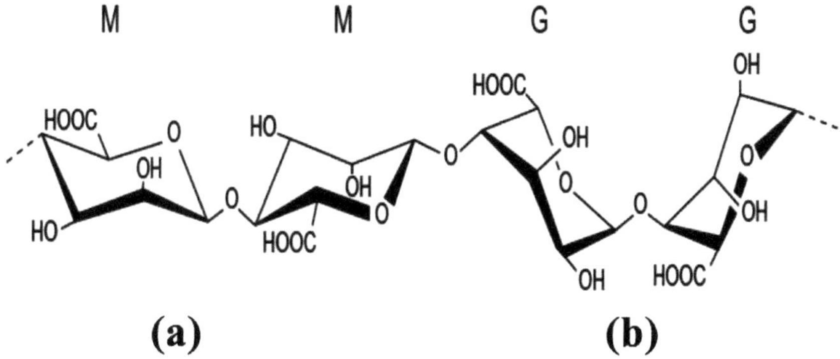

Fig. 2.3 Schematic representation of the structural units of alginate block types. **a** β-[1–4)-D-Mannuronic acid; **b** α-[1–4)-L-guluronic acid

Alginates

Seaweeds' flexibility can be attributed, at least in part, to alginates, which are made up of mannuronic and guluronic acid units (Fig. 2.3). Each monomer's carboxyl group gives the alginates an anionic character that varies with pH; as a result, the alginates can be extracted as either salt (known as alginates) or acid (known as alginic acids). Guuronic acid-based polymer segments have a coiled shape that creates cavities that can hold Ca^{2+} ions or other ions like Fe^{3+} and Al^{3+}. This stabilizes the structure and gives the alginates their gelling qualities, which are what make them so well-known in the marketplace. Alginates are used as auxiliary agents in the pharmaceutical and cosmetics industries to treat heartburn, esophagitis, and medication delivery as gelling agents, and in the food industry as a non-toxic additive [E400–405] (Lee and Mooney 2012; Kumar et al. 2017; Rhein-Knudsen et al. 2017).

Laminaran

Laminarin constitutes up to 35% of the dry weight of the brown algae and is a storage PS. The purified preparation is a group of low-molecular-weight (about 5 kDa) water-soluble PS and is a member of the 1–3-β-D-glucan family, consisting of a linear structure of [1,3)-β-D-glucopyranose residues [about 20–25 glucose moieties] with a small degree of branching (Fig. 2.4) (Salehi et al. 2019).

Agars

Red seaweeds mostly contain agars and carrageenans as structural polysaccharides. Agar refers to both agarose and agaropectin, with agarose accounting for up to 70% of the polysaccharide. Agaropectin has more substituents (sulfate, methoxy, and

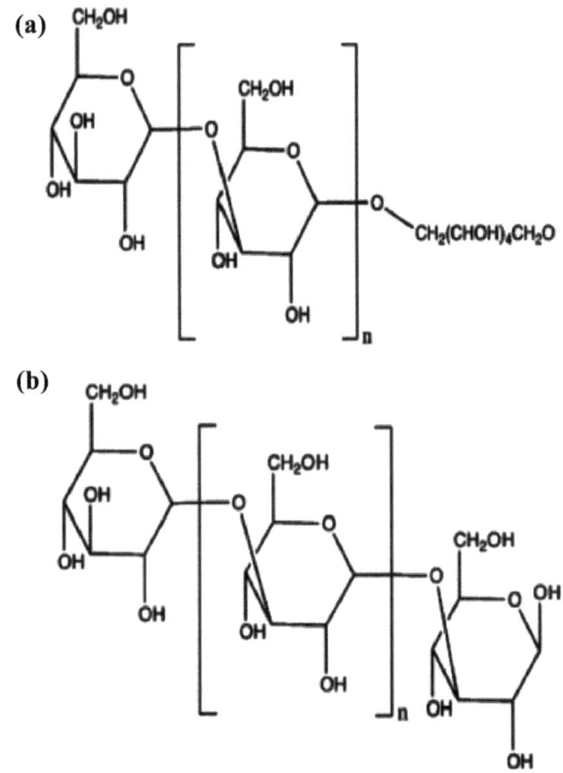

Fig. 2.4 Schematic representation of the structural units of Laminaran

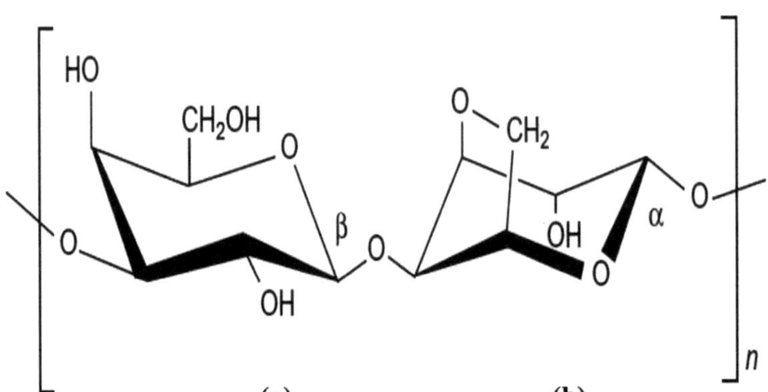

Fig. 2.5 Schematic representation of the structural units of the agarose block types. **a** D-Galactose; **b** 3,6-anhydro-L-galactose

pyruvate groups) at the C-2, C-4, and C-6 carbons of the _-d-galactose unit and the C-2, C-3, and C-6 of the other sugar monomer (Fig. 2.5) (Lee et al. 2017).

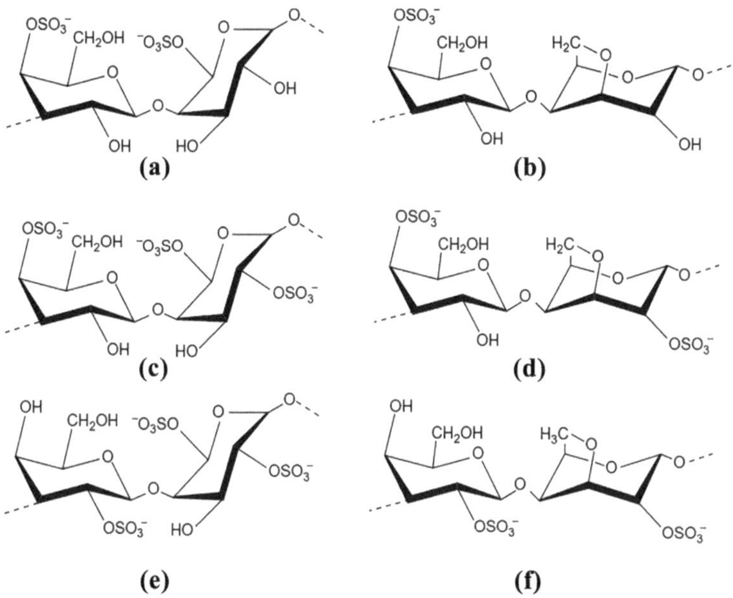

Fig. 2.6 Schematic representation of the chemical structures of different types of carrageenan: **a** μ-carrageenan; **b** κ-carrageenan; **c** ν-carrageenan; **d** ι-carrageenan; **e** λ-carrageenan; **f** θ-carrageenan

Carrageenans

The majority of red seaweed contains carrageenans, which are sulfate polysaccharides that the human digestive system cannot absorb. Due to their ability to form thermoreversible gels, these dietary fibers are highly valuable commercially. They are used in the food, pharmaceutical, and cosmetic industries for things like drug delivery, tissue engineering, and antivirals, as well as in dairy products as E407 or E407a. The use of carrageenan as an active ingredient in pharmaceuticals is still relatively new (Fig. 2.6) (Cunha and Grenha 2016).

Ulvans

Green seaweeds are the primary source of ulvans, a class of sulfated, water-soluble heteropolysaccharides with gelling qualities. Green algae have an ulvan content ranging from 8 to 29% [DW]. It is mostly made up of glucuronic acid, rhamnose, and xylose sulfated, with trace amounts of glucose and iduronic acids. Ulvans have a variety of biological properties, including anticoagulant, antioxidant, and immunomodulatory effects (Fig. 2.7) (Lahaye and Robic 2007).

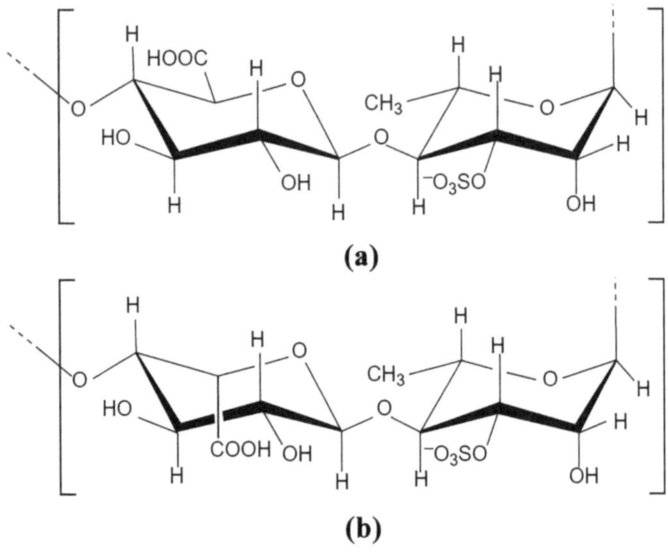

Fig. 2.7 Schematic representation of the repeating units of the ulvan structure. **a** Ulvanobiuronic acid type A3s, composed of β-D-glucuronic acid and α-L-rhamnose-3-sulfate; **b** ulvanobiuronic acid type B3s, composed of α-L-iduronic acid and α-L-rhamnose-3-sulfate

2.1.1.2 Protein

The third most prevalent class of primary metabolites is proteins. Compared to the green and brown seaweed species, the red seaweed species have higher protein levels (Rohani-Ghadikolaei et al. 2012). Red seaweeds are a viable substitute for terrestrial proteins because they share an essential amino acid profile with ovalbumen. Red seaweed is a desirable protein source due to its high levels of essential amino acids (EAAs) and utilizable protein (up to 47%). They typically have low calorie value but high protein and dietary fiber, along with vital vitamins, minerals, and bioactive components, albeit species and environment have an impact on the nutrients. Furthermore, red seaweed has been found to have a variety of possible bioactive peptides, indicating its promise as a cost-effective, nutrient-dense, and sustainable marine source (Ullah et al. 2023; Nowacka et al. 2023; Rawiwan et al. 2022). Proteins are polymers formed by the combination of different amino acids linked by an amide bond, with varying ratios of each amino acid and 20 distinct amino acids (Fig. 2.8) (Salehi et al. 2019).

2.1.1.3 Lipids

The term "lipids" refers to a class of substances that include fatty acids (saturated, mono-, di-, and polyunsaturated), waxes, acylglycerols (mono-, di-, and triacylglycerols), phospholipids, glycoglycerolipids, fat-soluble vitamins (such as vitamins A

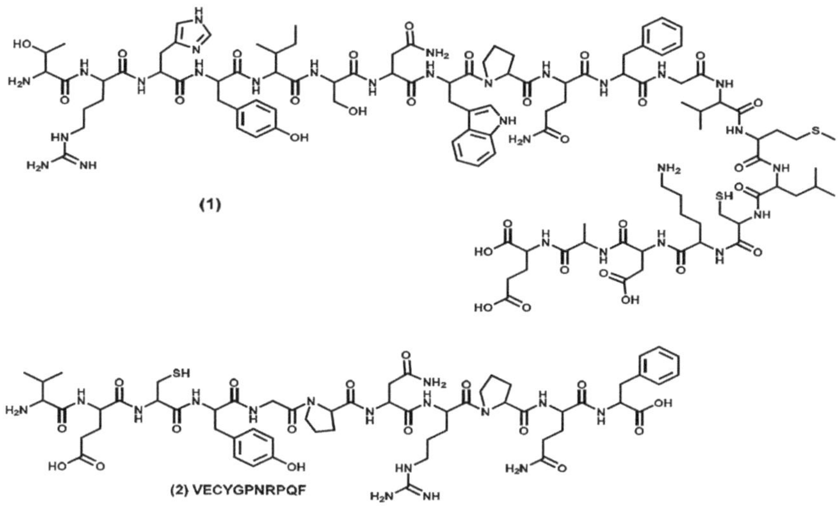

Fig. 2.8 Primary structure of a hypothetical protein consisting of all amino acids (1) and the seaweed peptide with potent antioxidant activity (2) (Salehi et al. 2019)

and E), sterols, and carotenoids. Only 1 to 5% of algal dry matter is made up of lipids, which have an intriguing composition of polyunsaturated fatty acids, especially omega-3 and omega-6 acids, which are crucial for preventing diabetes, osteoarthritis, and cardiovascular illnesses (Fig. 2.9) (Darcy-Vrillon 1993; Susanto et al. 2019).

2.1.1.4 Minerals

Although minerals [Ca, Fe, I, Mg, P, K, Zn, Cu, Mn, Se, and F] are essential to the life cycle of macroalgae, they are not biosynthesized by them, but rather absorbed from the environment, depending on several factors such as salinity, temperature, pH, and light. Because of their superior capacity for bioabsorption and bioaccumulation, macroalgae typically contain more minerals than terrestrial plants. The sea provides seaweeds with an incredible wealth of minerals, macro, and trace elements. It's interesting to note that the mineral component of some seaweeds can make up as much as 36% of their dry mass; this depends on the species, time of year, exposure to waves, physiological and environmental factors, and the techniques used for processing and mineralization (Mabeau and Fleurence 1993; Fleurence and Coeur 1993). According to El Zokm et al. (2021), compared to other categories, red seaweed has a greater average amount of Ca, S, B, Cl, Na, K, Li, Mn, Ni, Co, Cd, and Pb. On average, green seaweeds have higher amounts of I, Mg, Fe, and Cu than other seaweeds. The average levels of Br, F, Si, P, Cr, and Zn are higher in the brown seaweeds.

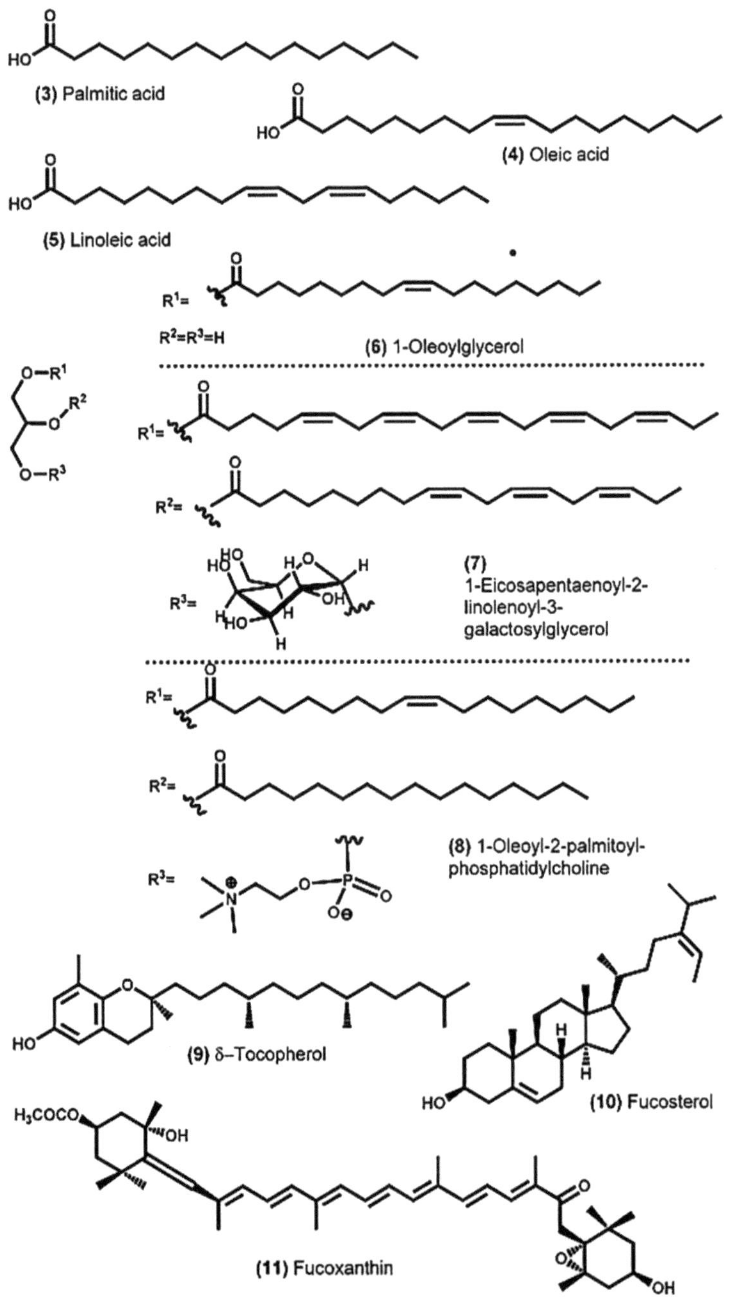

Fig. 2.9 Examples of the chemical structures of seaweed-found lipophilic vitamins (9), sterols (10), phospholipids (8), acylglycerols (6), saturated and unsaturated fatty acids (3–5), and carotenoids (11) (Salehi et al. 2019)

2.1.1.5 Pigments

Through photosynthesis, pigments are compounds that absorb light and convert it into energy. According to Salido et al. (2024), seaweeds are naturally colored by three main photosynthetic pigments: phycobilins, carotenoids, and chlorophylls. The process of photosynthesis requires chlorophylls, particularly chlorophyll a (Chl a). Porphyrin is the parent compound of chlorophyll. There is a magnesium atom in the core of chlorophyll's chemical structure. Additionally, the magnesium is surrounded by four nitrogen atoms. A pyrrole ring structure with four carbons and nitrogen is present. There's also a hydrocarbon end. Together with nitrogen and magnesium, the four pyrrole rings are referred to as the porphyrin head. Phytol tails are the hydrocarbon ends (Fig. 2.10) (Palmer et al. 2021).

Carotenoids and phycobilins are accessory pigments that conduct energy to Chl a. Macroalgae belonging to the same group have some similarities in the pigment

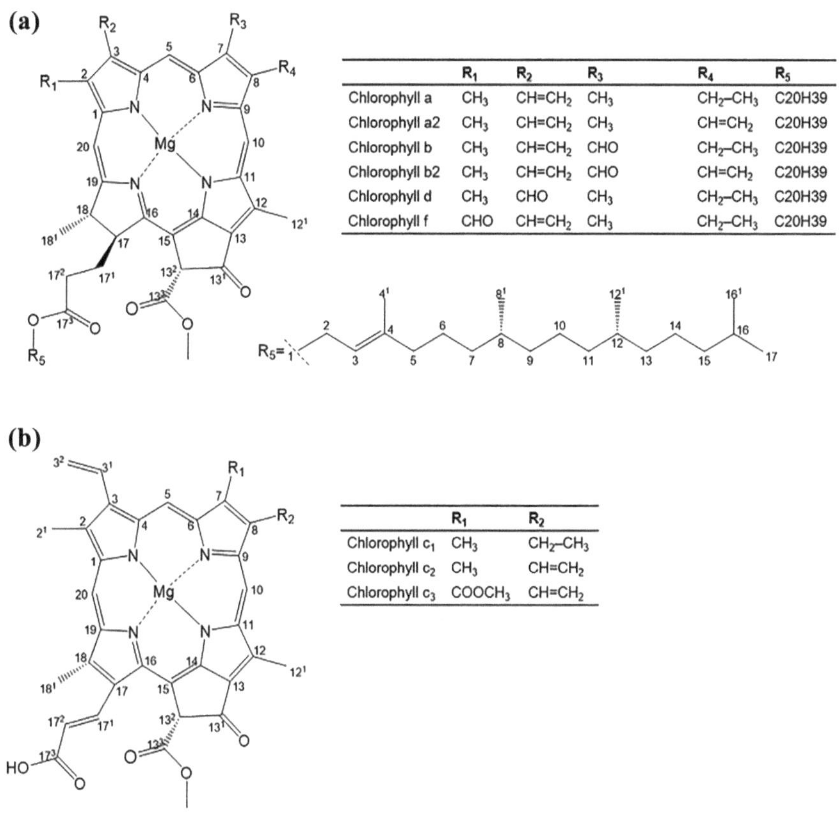

	R_1	R_2	R_3	R_4	R_5
Chlorophyll a	CH_3	$CH{=}CH_2$	CH_3	$CH_2{-}CH_3$	C20H39
Chlorophyll a2	CH_3	$CH{=}CH_2$	CH_3	$CH{=}CH_2$	C20H39
Chlorophyll b	CH_3	$CH{=}CH_2$	CHO	$CH_2{-}CH_3$	C20H39
Chlorophyll b2	CH_3	$CH{=}CH_2$	CHO	$CH{=}CH_2$	C20H39
Chlorophyll d	CH_3	CHO	CH_3	$CH_2{-}CH_3$	C20H39
Chlorophyll f	CHO	$CH{=}CH_2$	CH_3	$CH_2{-}CH_3$	C20H39

	R_1	R_2
Chlorophyll c_1	CH_3	$CH_2{-}CH_3$
Chlorophyll c_2	CH_3	$CH{=}CH_2$
Chlorophyll c_3	$COOCH_3$	$CH{=}CH_2$

Fig. 2.10 Structure of chlorophylls (Palmer et al. 2021)

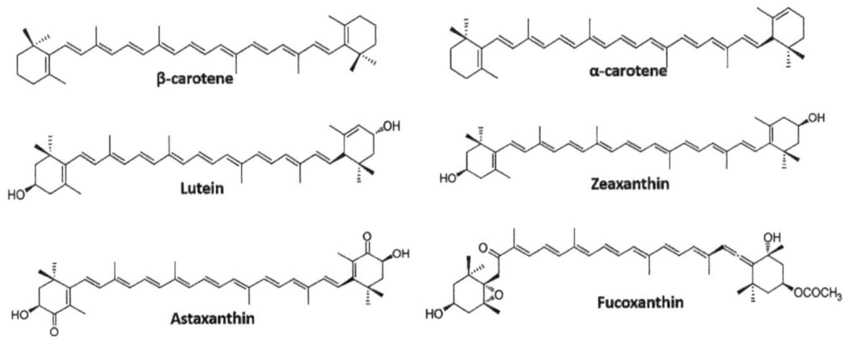

Fig. 2.11 Structures of major carotenoids in seaweed

profile. The highest chlorophyll quantities are found in green seaweed; brown macrophytes are characterized by their fucoxanthin content, while red ones are known for their phycobilins (Cikos et al. 2022).

Because they are more likely to accumulate under stressful conditions, macroalgal pigments can be used as markers of environmental stress. Despite their protective function, they might deteriorate under stress. The temperature, salinity, and acidification of the saltwater are some of the many variables that affect the pigment content of macroalgae, but each of these variables also depends on the particular macroalgal species (Cikos et al. 2022; Salido et al. 2024).

Carotenoids as Accessory Pigments in Photosynthesis

Carotenoids are the main accessory pigment found in algae. They belong to the tetraterpenoid family, contain 40 carbon atoms, and absorb light in the 400–500 nm region. Structurally, these compounds are composed of two small six-carbon rings connected by a polyene chain of carbon atoms. They are insoluble in water and are normally attached to proteins that are attached to the membrane. There are over 600 types of carotenoids, which are classified as either carotenes or xanthophylls.

Polyene chain can be substituted with cyclic groups and oxygen-containing functional groups in the biosynthesis of carotenoids. These oxygen-containing carotenoids are called xanthophylls. Hence, carotenoids were divided into two groups: (1) carotenes and (2) xanthophylls. Furthermore, xanthophylls were classified based on the nature of oxygen present, for example, lutein [oxygen is present as –OH], canthaxanthin [as oxy groups], and astaxanthin [a combination of –OH and oxy groups]. The carotenes are soluble in nonpolar solvents, whereas xanthophylls dissolve in polar solvents (Fig. 2.11) (Ratih 2018).

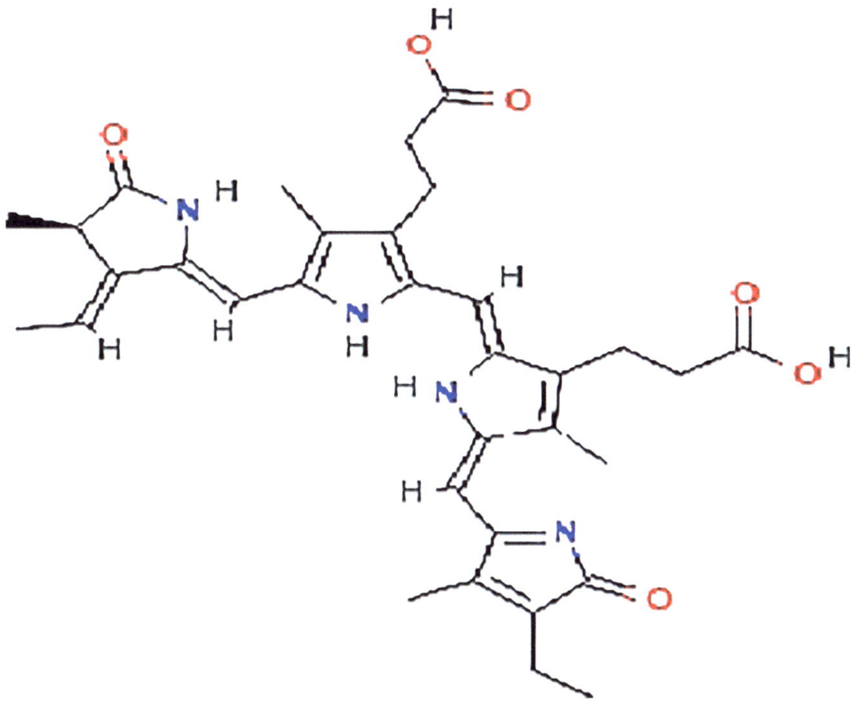

Fig. 2.12 Structures of phycobilins in seaweed

Phycobilins [PBs)

They are unique and vibrant pigments in various photosynthetic organisms, particularly in red algae. These pigments are crucial in capturing and transferring light energy for photosynthesis (Fig. 2.12). However, beyond their photoreceptive functions, PBs have garnered significant attention due to their diverse biological activities and potential applications in various fields (García-Gómez et al. 2025).

2.1.1.6 Vitamins

Vitamin C [ascorbic acid], B1 [thiamine], B2 [riboflavin], B3 [niacin], and B12 [cobalamin] are among the most common water-soluble vitamins found in seaweeds, according to Lovander et al. (2018), Fat-soluble vitamins like A and E have also been found (Fig. 2.13) (Cadar et al. 2025).

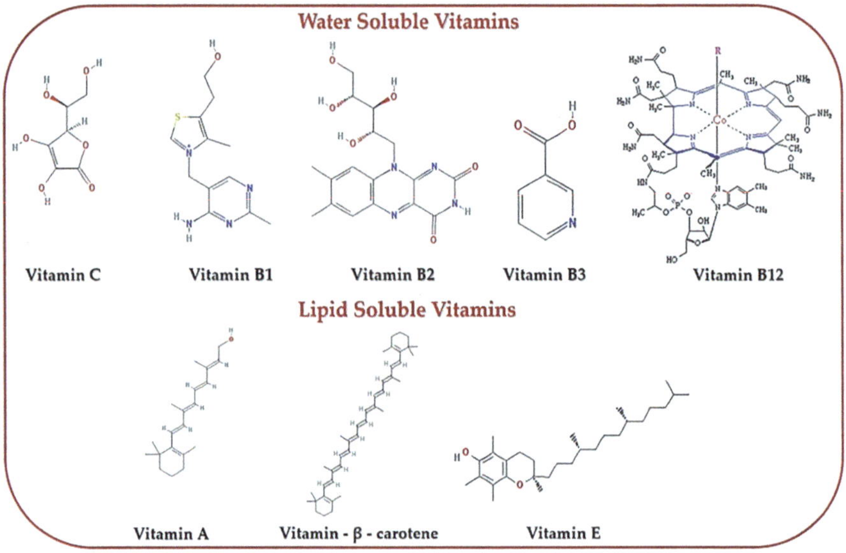

Fig. 2.13 Chemical structure of the most common vitamins in seaweeds (Cadar et al. 2025)

2.1.2 *Secondary Metabolites*

The main secondary metabolites produced in algae tissues are phenolic compounds, halogenated compounds, alkaloids, sterols, terpenes, and small peptides, among other bioactive compounds (Stengel et al. 2011; Holdt and Kraan 2011). Marine algae contain many types of alkaloids, including the indole, phenylethylamine, naphthyridine, and bis-indole types. Algae frequently include terpenoids such as monoterpenes, sesquiterpenes, diterpenes, triterpenes, and tetraterpenes, such as carotenoids. Algae also incorporate secondary metabolites such as glycosides, amines, steroids, and polyketides (Rocha et al. 2018).

2.1.2.1 Phenolic Compounds

One of the phytochemicals found in seaweeds is phenolic compounds, which can range in structure from simple molecules to high-molecular-weight molecules and are composed of an aromatic ring with one or more hydroxyl functional groups (Aina et al. 2022). The location of the hydroxy group, the quantity of hydroxyl groups, and the number of phenyl rings in the structure all affect the bioactivity of phenolic compounds (Aina et al. 2022). The most varied group of secondary metabolites and their derivatives, including polyphenolics, which include flavonoids, anthocyanins, and tannins, is found in the compound class made up of phenolic metabolites. Seaweeds include phenolics, including flavonoids, coumarins, polyphenols, and phlorotannins. According to Tsimogiannis and Oreopoulou (2019), lignins, the

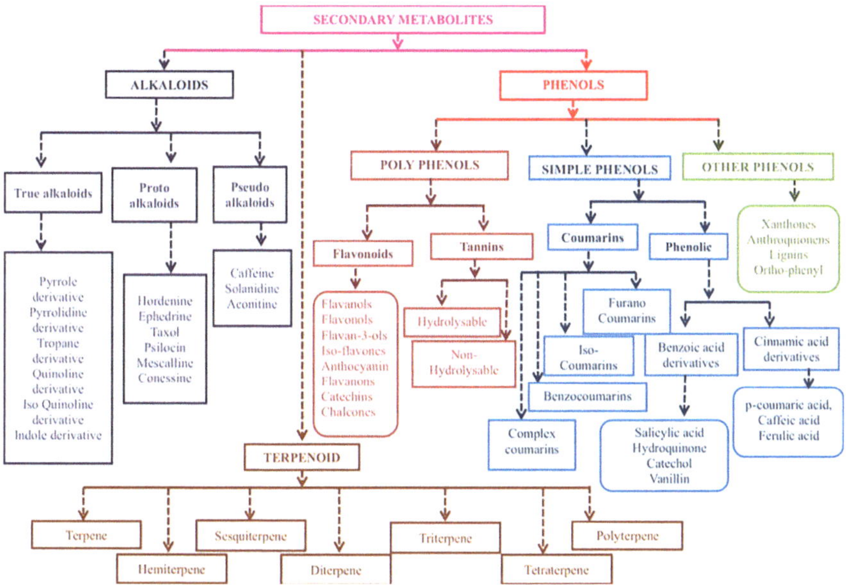

Fig. 2.14 Classification of the most common secondary metabolites in seaweed (Priyanka et al. 2022)

phenolic polymers, are very rare or less explored in algae (Fig. 2.14) (Priyanka et al. 2022).

2.1.2.2 Alkaloids

A chemical with nitrogen atoms in a cyclic ring is called an alkaloid. The term "alkaloid" encompasses a wide range of biological amines and compounds that contain halogenated cyclic nitrogen. The latter is unique to marine algae and organisms. Terrestrial plants have not been identified to contain them. Three categories of alkaloids were identified in marine algae: (i) phenylethylamine alkaloids; (ii) indole and halogenated indole alkaloids; and (iii) additional alkaloids, including derivatives of 2,7-naphthyridine. The majority of the alkaloids that have been extracted from marine algae are indole and 2-phenylethylamine families. Alkaloids that contain bromine and chloride are known as halogenated alkaloids, and they are specifically found in Chlorophyta. Rhodophyta contains the majority of the indole group's alkaloids (Fig. 2.14) (Barbosa et al. 2014; Priyanka et al. 2022).

2.1.2.3 Terpenes and Terpenoids

Based on an isoprene structure, terpenes (diterpenes, triterpenes, tetraterpenes, hemiterpenes, and sesquiterpenes) are created; when they have extra oxygen, they are called terpenoids. According to Priyanka et al. (2022), seaweeds contain a wide variety of terpenes and terpenoids with a range of biological activities (Fig. 2.14).

2.1.2.4 Lectins

Lectins are non-immune response glycoproteins that bind reversibly to polysaccharides with at least one non-catalytic domain, glycans of glycoproteins, and glycolipids, resulting in agglutination. Because they are monomeric, low molecular weight proteins with a high concentration of acidic amino acids, algae lectins are different from terrestrial lectins. The majority of them exhibit more specificity for oligosaccharides and/or glycoproteins than for monosaccharides, and they do not require metal ions for their biological activities. Algal lectins are classified into three main groups according to their ability to bind to glycoproteins: complex type N-glycan-specific lectins, high mannose (HM) type N-glycan-specific lectins, and lectins that are specific to both of the aforementioned N-glycan types. Additionally, lectins from marine species are divided into rhamnose-binding lectins, galectins, intelectins, F-type lectins, and C-type lectins (Fig. 2.15) (Cheung et al. 2015; Singh et al. 2015).

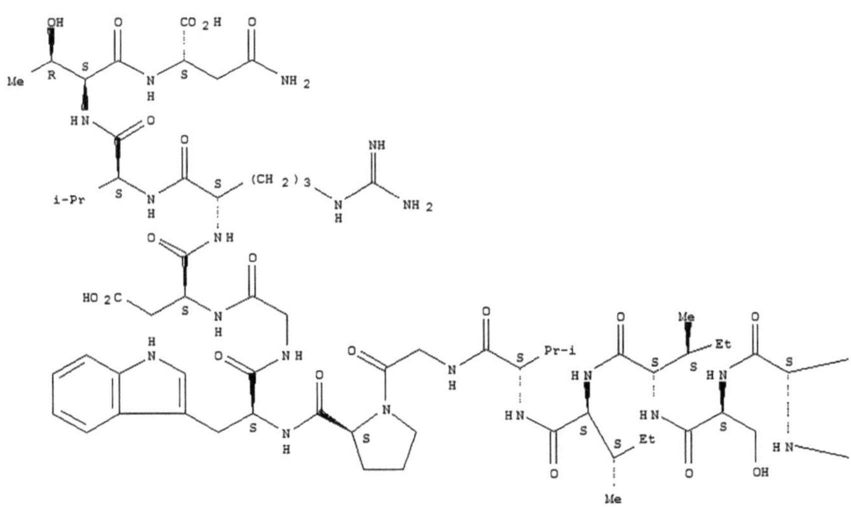

Fig. 2.15 Chemical structure of lectin

2.2 Omics Technologies

An "**Omics**" branch called "metabolomics" describes the thorough qualitative and quantitative examination of all (or a subset of) metabolites in a biological system at a specific moment in time (Tanna and Mishra 2018). Metabolomics has been defined as "the quantitative measurement of the dynamic multiparametric response of a living system to pathophysiological stimuli or genetic modification" [1]. Metabolomics has emerged as a powerful tool for gaining comprehensive insights into the biochemical composition of living organisms (Farag 2014; Noreldeen et al. 2025). Metabolomics can be used to explore the impact of geography on bioactivity, the impact of storage conditions, and even an organism's response to diverse environmental stressors. More than 15,000 primary and secondary metabolites from distinct macroalgal metabolic pathways have been found, with applications spanning from food and feed to medicine and wastewater treatment, biofertilizer, and biofuel. Awareness of defense responses has grown fast as a result of novel omics methods and genome-editing tools. High-throughput "**Omics**" technologies have rapidly advanced our knowledge of edible seaweeds, unraveling the complexity, functions, and interactions of biomolecules: DNA [genomics], RNA [transcriptomics], proteins [proteomics) and metabolites (metabolomics) have advanced our understanding of cellular biology and discovered critical components in metabolic networks (Xu et al. 2024; Varadharajan et al. 2025).

Omics approach enables the simultaneous detection of hundreds of metabolites, facilitating comparisons between samples to identify differences and similarities in a semi-automated, untargeted manner. Metabolomics typically employs chromatography techniques such as gas chromatography (GC) or liquid chromatography (LC), often coupled with mass spectrometry (MS) or nuclear magnetic resonance spectroscopy (NMR), to analyze complex metabolite mixtures. While NMR provides a holistic metabolite profile, LC-MS offers superior resolution of individual chemical components, enabling the detection of novel low-abundance metabolites (Zhang et al. 2012; Gika et al. 2014). Classical biochemical techniques are time-consuming and inefficient, whereas global and high-throughput analytical techniques are the basis of omics technologies, such as microarray, 2D-gel, and 2DLC/MS, which generate large-scale data. Omics is expected to provide many additional insights and clues to the mechanisms of biological processes and functions via bioinformatics and computer modelling, perhaps establishing a theoretical framework for modern biology. Using a variety of computational and informatics frameworks, targeted "omics" research on seaweeds has the potential to quickly advance our knowledge of molecular biological processes and help address the most important environmental and social problems of our day (Figs. 2.16 and 2.17) (Patwary et al. 2021).

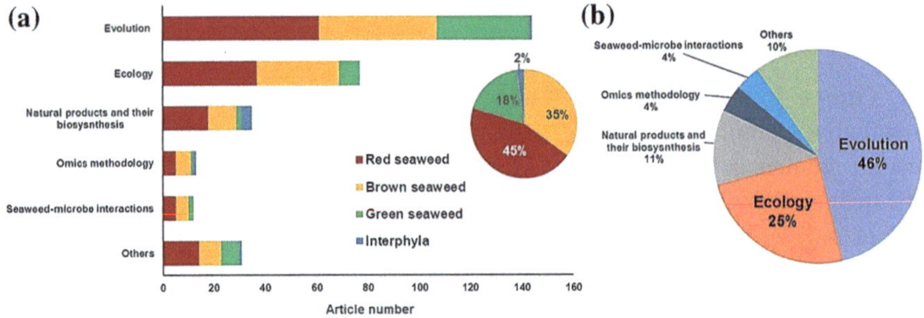

Fig. 2.16 Overview of seaweed omics research. **a** Bar graph showing the number of articles in different research categories and within each seaweed group. Inset, pie chart showing the proportion of articles in each seaweed group. **b** Pie chart showing the proportion of the six main research categories in seaweed omics (Patwary et al. 2021)

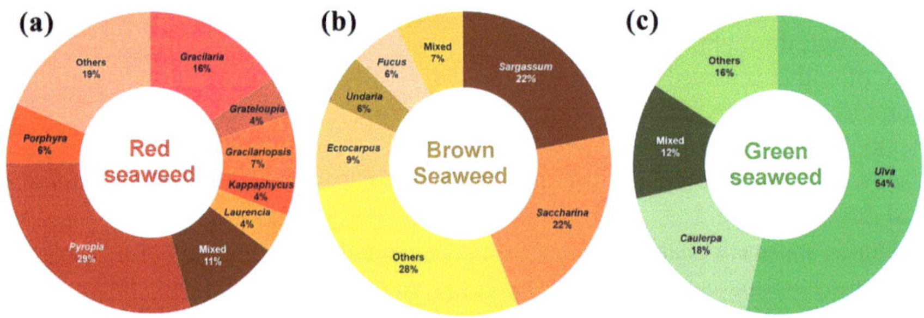

Fig. 2.17 Percentage of omics research articles derived from different genera. **a** Red seaweed genera. **b** Brown seaweed genera. **c** Green seaweed genera (Patwary et al. 2021)

2.2.1 Case Study for Omics Technologies (Noreldeen et al. 2025)

2.2.1.1 Energy and Carbohydrate Metabolism

The quantity of metabolites from the sugar alcohol metabolism, the Tricarboxylic Acid (TCA) cycle, and gluconeogenesis/glycolysis was shown to vary. Except for glucose, which was more prevalent in green and brown algae, the amounts of the majority of gluconeogenesis/glycolysis metabolites were similar throughout the seaweeds under investigation. Citrate, cis-aconitate, isocitrate, and succinate metabolites were among the many TCA cycle intermediates that were considerably ($p < 0.05$) lower in red algae and highly concentrated in brown algae (Fig. 2.18). Schema of reaction: Greater relative abundance in brown, red, and green macroalgal species is indicated by metabolites that are written in those colors, respectively. Black metabolites show no discernible variation between samples, while gray metabolites show values below the detection threshold.

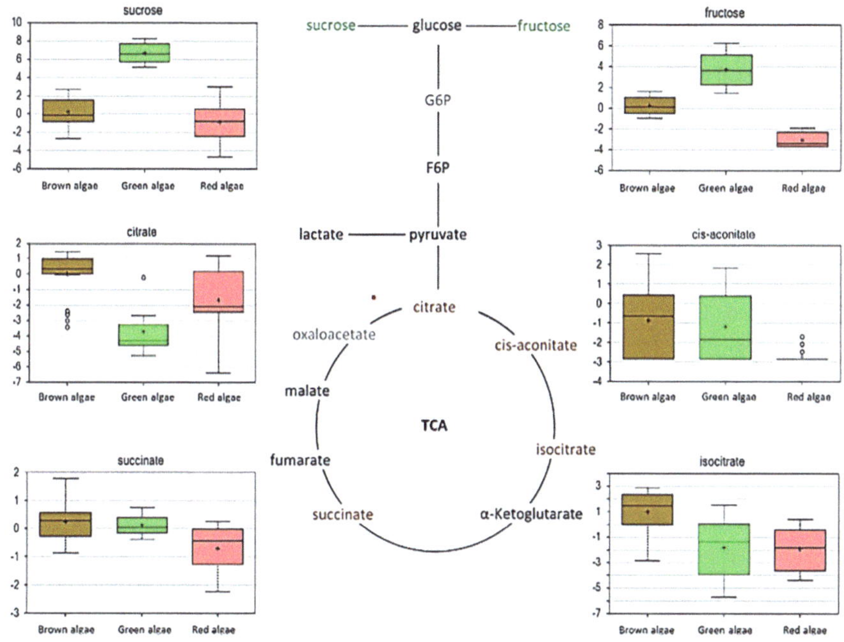

Fig. 2.18 Differences in the metabolites involved in energy metabolism between red, green, and brown algal samples (Belghit et al. 2017)

Box plots: Red, green, and brown algae species are shown by the x-axis, respectively. The scaled intensity mean (+) and median (—) values are displayed in the y-axis box plots. The upper and lower quartiles are represented by the top and bottom ranges of boxes, respectively, while the maximum and minimum distribution of the data are shown by the top and bottom whiskers. Fructose-6-phosphate (F6P) and glucose-6-phosphate (G6P).

References

Aina O, Bakare OO, Daniel AI, Gokul A, Beukes DR, Fadaka AO, Keyster M, Klein A (2022) Seaweed-derived phenolic compounds in growth promotion and stress alleviation in plants. Life 12:1548. https://doi.org/10.3390/life12101548

Akter A, Khairul M, Sobuj A, Islam S, Chakroborty K, Tasnim N, Ayon MH, Hossain F, Rafiquzzaman SM (2024) Seaweed polysaccharides: sources, structure and biomedical applications with special emphasis on antiviral potentials. Futur Foods 10:100440

Balboa EM, Gallego-Fábrega C, Moure A, Domínguez H (2016) Study of the seasonal variation on proximate composition of oven-dried Sargassum muticum biomass collected in Vigo Ria, Spain. J Appl Phycol 28:1943–1953

Barbosa M, Valentão P, Andrade PB (2014) Bioactive compounds from macroalgae in the new millennium: implications for neurodegenerative diseases. Mar Drugs 12:4934–4972

Belghit I, Rasinger JD, Heesch S, Biancarosa I, Liland N, Torstensen B, Waagbø R, Lock E-J, Bruckner CG (2017) In-depth metabolic profiling of marine macroalgae confirms strong biochemical differences between brown, red and green algae. Algal Res 26:240–249. https://doi.org/10.1016/j.algal.2017.08.001

Cadar E, Popescu A, Dragan A-M-L, Pesterau A-M, Pascale C, Anuta V, Prasacu I, Velescu BS, Tomescu CL, Bogdan-Andreescu CF et al (2025) Bioactive compounds of marine algae and their potential health and nutraceutical applications: a review. Mar Drugs 23:152. https://doi.org/10.3390/md23040152

Carpena M, Garcia-Perez P, Garcia-Oliveira P et al (2023) Biological properties and potential of compounds extracted from red seaweeds. Phytochem Rev 22:1509–1540. https://doi.org/10.1007/s11101-022-09826-z

Cavalier-Smith T (2000) Membrane heredity and early chloroplast evolution. Trends Plant Sci 5:174–182. https://doi.org/10.1016/S1360-1385[00)01598-3

Chan CX, Bhattacharya D (2010) The origin of plastids. Nativ Educ 3(9):84

Cheung RCF, Wong JH, Pan W, Chan YS, Yin C, Dan X, Ng TB (2015) Marine lectins and their medicinal applications. Appl Microbiol Biotechnol 99:3755–3773

Cikoš AM, Šubarić D, Roje M, Babić J, Jerković I, Jokić S (2022) Recent advances on macroalgal pigments and their biological activities (2016–2021). Algal Res 65:102748. https://doi.org/10.1016/j.algal.2022.102748

Cunha L, Grenha A (2016) Sulfated seaweed polysaccharides as multifunctional materials in drug delivery applications. Mar Drugs 14:42

Darcy-Vrillon B (1993) Nutritional aspects of the developing use of marine macroalgae for the human food industry. Int J Food Sci Nutr 44:23–35

El Zokm GM, Ismail MM, El-Said GF (2021) Halogen content relative to the chemical and biochemical composition of fifteen marine macro and micro algae: nutritional value, energy supply, antioxidant potency, and health risk assessment. Environ Sci Pollut Res 28:14893–14908. https://doi.org/10.1007/s11356-020-11596-0

El Zokm GM, Ismail MM, El Okba M (2022) Seaweed as a bioindicator of organic micropollutants polycyclic aromatic hydrocarbons (PAHs), and organochlorine pesticides (OCPs). Environ Sci Pollut Res. https://doi.org/10.1007/s11356-022-18634-z

Fleurence J, Coeur C (1993) The Influence of mineralization methods on the determination of the mineral content of brown seaweed Undaria pinnatifida by atomic absorption spectrophotometry. Hydrobiologia 260–261:531–534

Farag MA (2014) Comparative mass spectrometry & nuclear magnetic resonance metabolomic approaches for nutraceuticals quality control analysis: a brief review. J Recent Pat Biotechnol 8(1):17–24

García-Gómez C, Aguirre-Cavazos DE, Chávez-Montes A, Ballesteros-Torres JM, Orozco-Flores AA, Reyna-Martínez R, Torres-Hernández ÁD, González-Meza GM, Castillo-Hernández SL, Gloria-Garza MA, Kačániová M, Ireneusz-Kluz M, Elizondo-Luevano JH (2025) Phycobilins versatile pigments with wide-ranging applications: exploring their uses, biological activities, extraction methods and future perspectives. Mar Drugs 23(5):201. https://doi.org/10.3390/md23050201

Gika HG, Wilson ID, Theodoridis GA (2014) LC–MS-based holistic metabolic profiling. Problems, limitations, advantages, and future perspectives. J Chromatogr B 966:1–6

Holdt SL, Kraan S (2011) Bioactive compounds in seaweed: functional food applications and legislation. J Appl Phycol 23:543–597

Islam MS, Sobuj MKA, Islam HMR, Hosain ME, Rashid MH (2022) Present status of Seaweed resources in Bangladesh: a review on the diversity, culture methods and utilization. Bangladesh J Zool 50:283–307. https://doi.org/10.3329/bjz.v50i3.65537

Khairy HM, El-Shafay SM (2013) Seasonal variations in the biochemical composition of some common seaweed species from the coast of Abu Qir Bay, Alexandria, Egypt. Oceanologia 55:435–452

Kumar V, Latha S, Rose MH, Sudha PN (2017) Industrial applications of alginate. In: Sudha PN (ed) Industrial applications of marine biopolymers. CRC Press, Boca Raton, FL, USA, pp 545–575
Lahaye M, Robic A (2007) Structure and functional properties of ulvan, a polysaccharide from green seaweeds. Biomacromol 8:1765–1774
Lee W-K, Lim Y-Y, Leow AT-C, Namasivayam P, Abdullah JO, Ho C-L (2017) Biosynthesis of agar in red seaweeds: a review. Carbohydr Polym 164:23–30
Lee KY, Mooney DJ (2012) Alginate: properties and biomedical applications. Prog Polym Sci 37:106–126
Lovander MD, Lyon JD, Parr DL, Wang J, Parke B, Leddy J (2018) Critical review—electrochemical properties of 13 Vitamins: a, critical review and assessment. J Electrochem Soc 165:18–49
Mabeau S, Fleurence J (1993) Seaweed in food products: biochemical and nutritional aspects. Trends Food Sci Technol 4:103–107
Marinho-Soriano E, Fonseca PC, Carneiro MAA, Moreira WSC (2006) Seasonal variation in the chemical composition of two tropical seaweeds. Bioresour Technol 97:2402–2406
Noreldeen HAA, Hamed AM, El-Shazly M, El-Saharty AA, Farghaly OA, Huang S (2025) Integrating untargeted metabolomics and computational docking for biomarker evaluation: a case study on marine algae-derived ligands. Bioorg Chem 161:108539. https://doi.org/10.1016/j.bioorg.2025.108539
Nowacka M, Trusinska M, Chraniuk P, Drudi F, Lukasiewicz J, Nguyen NP, Przybyszewska A, Pobiega K, Tappi S, Tylewicz U et al (2023) Developments in plant proteins production for meat and fish analogues. Molecules 28:2966. https://doi.org/10.3390/molecules28072966
Padam BS (2020) Sustainable seaweed technologies||Seaweed components, properties, and applications, pp 33–87. https://doi.org/10.1016/B978-0-12-817943-7.00002-0
Palmer JS, Muñoz DC, Edwards C (2021) Biological activities of natural products from microalgae. Cult Microalgae Food Ind. https://doi.org/10.1016/B978-0-12-821080-2.00004-6
Park E, Yu H, Lim J-H, Choi J, Park K-J, Lee J (2023) Seaweed metabolomics: a review on its nutrients, bioactive compounds and changes in climate change. Food Res Int 163:112221. https://doi.org/10.1016/j.foodres.2022.112221
Patwary ZP, Paul NA, Nishitsuji K, Campbell AH, Shoguchi E, Zhao M, Cummins SF (2021) Application of omics research in seaweeds with a focus on red seaweeds. Brief Funct Genomics 20:148–161. https://doi.org/10.1093/bfgp/elab023
Premarathna AD, Tuvikene R, Fernando PHP et al (2022) Comparative analysis of proximate compositions, mineral and functional chemical groups of 15 different seaweed species. Sci Rep 12:19610. https://doi.org/10.1038/s41598-022-23609-8
Priyanka KR, Rajaram R, Sivakumar SR (2022) A critical review of the pharmacological properties of marine macroalgae. Biomass Conv Bioref. https://doi.org/10.1007/s13399-022-03134-4
Ratih P (2018) Bioactive seaweeds for food applications||Seaweed-derived carotenoids, pp 95–107. https://doi.org/10.1016/B978-0-12-813312-5.00005-4
Rawiwan P, Peng Y, Paramayuda IGPB, Quek SY (2022) Red seaweed: a promising alternative protein source for global food sustainability. Trends Food Sci Technol 123:37–56
Rhein-Knudsen N, Ale MT, Ajalloueian F, Meyer AS (2017) Characterization of alginates from Ghanaian brown seaweeds: Sargassum spp. and Padina spp. Food Hydrocoll 71:236–244
Rocha DHA, Seca AML, Pinto DCGA (2018) Seaweed secondary metabolites in vitro and in vivo anticancer activity. Mar Drugs 16(11):410. https://doi.org/10.3390/md16110410
Rohani-Ghadikolaei K, Abdulalian E, Ng WK (2012) Evaluation of the proximate, fatty acid and mineral composition of representative green, brown and red seaweeds from the Persian Gulf of Iran as potential food and feed resources. J Food Sci Technol 49(6):774–780. https://doi.org/10.1007/s13197-010-0220-0
Rosa GP, Tavares WR, Sousa PMC, Pagès AK, Seca AML, Pinto DCGA (2020) Seaweed secondary metabolites with beneficial health effects: an overview of successes in in vivo studies and clinical trials. Mar Drugs 18(1):8. https://doi.org/10.3390/md18010008

Salehi B, Sharifi-Rad J, Seca AML, Pinto DCGA, Michalak I, Trincone A, Mishra AP, Nigam M, Zam W, Martins N (2019) Current trends on seaweeds: looking at chemical composition, phytopharmacology, and cosmetic applications. Molecules 24(22):4182. https://doi.org/10.3390/molecules24224182

Salido M, Soto M, Seoane S (2024) Seaweed: Nutritional and gastronomic perspective. A review. Algal Res 77:103357. https://doi.org/10.1016/j.algal.2023.103357

Sánchez-Machado DI, López-Cervantes J, López-Hernández J, Paseiro-Losada P (2004) Fatty acids, total lipid, protein, and ash contents of processed edible seaweeds. Food Chem 85:439–444. https://doi.org/10.1016/j.foodchem.2003.08.001

Seca AML, Gouveia VLM, Barreto C, Silva AMS, Pinto DCGA (2018) Comparative study by GC-MS and chemometrics on the chemical and nutritional profile of Fucus spiralis L. juvenile and mature life-cycle phases. J Appl Phycol 30:2539–2548

Singh RS, Thakur SR, Bansal P (2015) Algal lectins as promising biomolecules for biomedical research. Crit Rev Microbiol 41(1):77–88. https://doi.org/10.3109/1040841X.2013.798780

Stengel DB, Connan S, Popper ZA (2011) Algal chemo-diversity and bioactivity: sources of natural variability and implications for commercial application. Biotechnol Adv 29:483–501

Susanto E, Fahmi AS, Hosokawa M, Miyashita K (2019) Variation in lipid components from 15 species of tropical and temperate seaweeds. Mar Drugs 17:630

Tanna B, Mishra A (2018) Metabolites unravel the nutraceutical potential of edible seaweeds: an emerging source of functional food. Compr Rev Food Sci Food Saf 17:1613–1624

Tsimogiannis D, Oreopoulou V (2019) Classification of phenolic compounds in plants. In: Polyphenols in plants. Elsevier, pp 263–284

Ullah MR, Akhter M, Khan ABS, Hasan MM, Bosu A, Yasmin F, Haque MA, Islam MA, Mahmud Y (2023) Seaweed: a prominent source of protein and other nutrients. Sustain Aquat Res 2(2):145–166

Usov AI, Bilan MI, Ustyuzhanina NE, Nifantiev NE (2022) Fucoidans of brown algae: comparison of sulfated polysaccharides from *fucus vesiculosus* and *Ascophyllum nodosum*. Mar Drugs 20(10):638. https://doi.org/10.3390/md20100638

Varadharajan V, Rajendran R, Muthuramalingam P, Runthala A, Madhesh V, Swaminathan G, Murugan P, Srinivasan H, Park Y, Shin H, Ramesh M (2025) Multi-omics approaches against abiotic and biotic stress—a review. Plants 14(6):865. https://doi.org/10.3390/plants14060865

Xie C, Lee ZJ, Ye S, Barrow CJ, Dunshea FR, Suleria HAR (2023) A review on seaweeds and seaweed-derived polysaccharides: nutrition, chemistry, bioactivities, and applications. Food Rev Intl 40(5):1312–1347. https://doi.org/10.1080/87559129.2023.2212055

Xu M, Zhang Y, Wu B et al (2024) A critical review of the edible seaweed palmaria palmata (L.) Weber & Mohr and its bioactive compounds in the "omics" era. Algal Res 82:103606. https://doi.org/10.1016/j.algal.2024.103606

Zhang A, Sun H, Wang P, Han Y, Wang X (2012) Modern analytical techniques in metabolomics analysis. J Therm Anal 137(2):293–300

Chapter 3
Macroalgal Applications

3.1 Ecological Application

Macroalgae provided multiple environmental benefits, including shoreline protection, habitats for various biota, eutrophication mitigation, ocean acidification modification, monitoring of pollutant, and water treatment. They serve a critical function in mitigating coastal marine pollution and in sequestering atmospheric CO_2, thereby contributing to the reduction of anthropogenic climate change (Cotas et al. 2023).

3.1.1 Co-benefits of Seaweeds and Ocean Afforestation

Due to the rapid economic growth in emerging economies, anthropogenic greenhouse gas emissions are expected to increase dramatically in the near future, exacerbating the issue. Consequently, it is essential to implement all possible strategies to reduce atmospheric CO_2 levels and prevent further ecological damage (Jhariya et al. 2021).

In response to climate change, a "blue carbon" paradigm has emerged, promoting the use of aquatic environments for sustainable food and fuel production through processes such as carbon capture, sequestration, and deep-sea carbon deposition (Yong et al. 2022). One promising approach within this framework is ocean afforestation, which involves large-scale seaweed farming in the open ocean to remove carbon dioxide. This method relies on creating expansive underwater forests of macroalgae, such as kelp and other seaweed species (Boyd et al. 2022).

Seaweed aquaculture plays a critical role in the coastal carbon cycle by functioning as an important biological carbon sink (Cotas et al. 2023), playing a key role in carbon sequestration, with an estimated capacity of 29.25 tonnes of carbon per year (Hossain et al. 2021). Beyond direct sequestration, macroalgae contribute to climate change mitigation through two additional mechanisms: first, by enabling carbon offsets—using macroalgal products to reduce industrial CO_2 emissions and second,

G. M. El Zokm et al., *Rethinking Seaweeds via Applications and Technologies*,
Environmental Science and Engineering, https://doi.org/10.1007/978-3-032-14956-5_3

by lowering methane emissions in agriculture through the use of algal-based livestock feed (Mrčelić et al. 2024).

Furthermore, seaweed aquaculture has been increasingly recognized as a strategic contributor to the achievement of the United Nations Sustainable Development Goals (SDGs) (Duarte et al. 2021). Ocean afforestation has been proposed as a promising carbon dioxide removal (CDR) strategy. Within this context, large-scale offshore seaweed cultivation—beyond continental shelf boundaries has been suggested as a mechanism for substantial oceanic carbon sequestration (Macreadie et al. 2019). Between 61 and 268 megatons of carbon can be restored by seaweeds annually, with an average of approximately 173 megatons. About 10% of this carbon is retained by being buried in coastal sediments, while the remaining 90% is transferred to the deep ocean waters (Farghali et al. 2023). According to Froehlich et al. (2019), up to 48 million km^2 of the global ocean surface equivalent to roughly 11% of the ocean area could be suitable for seaweed cultivation. Under theoretical maximum conditions, this could facilitate annual carbon sequestration of up to 72 Gigatonnes, surpassing current global CO_2 emissions, based on an upper-bound sequestration potential of 1500 tonnes of CO_2 per km^2 (Duarte et al. 2017).

Recently, Wang et al. (2023) suggested that the net effectiveness of ocean afforestation may be reduced by 20–100% due to two biogeochemical feedbacks mechanisms, including calcification by epibionts and nutrient redistribution effects.

Additional carbon sequestration pathways associated with seaweed farming include biofuel production, which reduces CO_2 emissions by substituting fossil fuels; use as biofertilizers, thereby decreasing reliance on synthetic fertilizers; methane emission reduction when incorporated into ruminant diets; and physical ecosystem services such as attenuating wave energy and protecting coastal zones, which contribute to climate change mitigation. Moreover, seaweed cultivation can enhance water pH and oxygen levels, thereby alleviating ocean deoxygenation and acidification (Farghali et al. 2023).

Ocean afforestation remains an emerging concept, characterized by significant research gaps. Consequently, increased investment in scientific investigation is necessary before it can be rigorously evaluated alongside other ocean-based climate mitigation approaches. The relatively limited scientific progress in ocean afforestation compared to the level of industry interest may be attributed to uncertainties surrounding high operational costs, engineering challenges, and the harsh environmental conditions associated with offshore aquaculture sites, which are the primary locations proposed for ocean afforestation (Fernand et al. 2017). Additionally, investors have historically favored nearshore seaweed aquaculture due to its established consumer markets and the potential for higher financial returns compared to cultivation solely aimed at carbon sequestration (van den Burg et al. 2016).

To evaluate the ecological viability of ocean afforestation utilizing seaweed, it must be taken into consider various factors including algal species selection, interspecific competition, and physicochemical parameters such as nutrient availability, irradiance, temperature, ocean currents, substrate types, salinity, and grazing pressures. These factors collectively influence algal growth and carbon sequestration potential (Ross et al. 2022).

Brown algae are considered prime candidates for ocean afforestation due to their greater carbon sequestration potential per unit area of growth substrate during cultivation cycles, attributable to their larger size relative to other macroalgal groups. These algae predominantly belong to two orders: Faceless (fucoid) and Laminariales (kelps), both of which are widespread habitat-forming species (Nelson 2020). Carbon stored by fucoid algae has a higher chance of being naturally transported and deposited in the deep ocean because they degrade at a slower rate compared to kelp species. This slower decomposition is linked to their higher content of structural tissues and secondary metabolites, which can deter grazers and decomposers (Birkemeyer et al. 2020). Conversely, kelps exhibit faster growth rates and can form large, dense populations, making them significant contributors to blue carbon sequestration as well (Krause-Jensen and Duarte 2016).

In addition to the hazards that need to be carefully considered, seaweed ocean afforestation offers a number of possible co-benefits. Research supporting co-benefits from seaweed farming has been primarily focused on near shore seaweed aquaculture; it is unclear whether these benefits would extend offshore. Co-benefits of ocean afforestation could include: (1) providing habitat for macrofaunal communities and potentially increasing fisheries productivity (Theuerkauf et al. 2021); (2) mitigating nutrient levels, especially in regions impacted by eutrophication or excessive anthropogenic nutrient inputs, by competing with photosynthetic microalgae that cause hypoxic conditions or by absorbing nutrients (Xiao et al. 2017; Visch et al. 2020); (3) lowering ocean acidification locally, can protect surrounding calcifying organisms (Xiao et al. 2021); (4) promoting phytoplankton carbon sequestration, is important, as lower acidification may boost local carbon fixation, which is otherwise inhibited at low pH levels (Petrou et al. 2019); and (5) creating jobs and promoting wider socioeconomic development can be achieved by establishing an ocean afforestation industry, which could benefit coastal communities (Larson et al. 2021).

Conversely, historically, the ecological and biosecurity issues associated with maritime afforestation have received relatively little attention (Russell et al. 2012). These effects will also be mostly scale-dependent, as demonstrated recently by Boyd et al. (2022), and they might be challenging to measure until large-scale ocean afforestation is carried out. The following are just a few of these effects: (1) phytoplankton and seaweed allelopathy or chemical competition, which may change offshore ecological functions (Boyd et al. 2022); (2) increased organic matter accumulation from the breakdown of seaweed biomass, which may cause oxygen depletion or hypoxic conditions in nearby waters (Wyrtki 1962); (3) Seaweeds release dissolved organic carbon (DOC), which can alter the quality and composition of DOC in the offshore environment. This could potentially affect the behavior of microbial communities, particularly oceanic heterotrophic bacteria, although the full extent of these effects is not yet fully understood (Boyd et al. 2022). (4) There is concern about the possibility of methane emissions from deep-sea microbial processes. Certain microorganisms in the deep ocean have the ability to convert deposited carbon into methane, which could contribute to the emission of greenhouse gases (Kotelnikova, 2002); and (5) there is a risk of marine life becoming entangled, including seabirds and marine mammals, similar to incidents observed in other forms of aquaculture (Young 2015).

A number of techniques have been proposed for growing seaweed as a method to mitigate carbon emissions (Froehlich et al. 2019). However, most of these suggestions are still in the early-stage commercial development and lack empirical validation or support in peer reviewed literature. The main challenge in efficiently sequestering carbon lies in the technological complexity of the process. For carbon derived from seaweed to be considered stored, the biomass must be deposited at significant ocean approximately 2000 m) (Baker et al. 2022). Various technological concepts have been introduced to address this challenge, which can be broadly categorized into three categories for technical ocean afforestation:

1. Traditional near shore aquaculture involves cultivating seaweed along the coast and then transporting it offshore for deposition in the deep ocean (Chung et al. 2017).
2. Systems for tethered offshore agriculture are established in deep-sea environments.
3. Non-tethered growth platforms are designed to operate independently or with the assistance of automated systems in the deep ocean.

Several variations of technical design themes that may be used in concert with these three technical categories. A selection of these conceptual themes are listed here and possible concepts are illustrated in Fig. 3.1. It is important to emphasize that these are just conceptual designs, and there is a much needed cost–benefit analysis of each of these preliminary concepts before application in the field. Attachment to existing infrastructure, for example, oil and gas or marine wind farms (van den Burg et al. 2020). Additionally, hybrid cultivation strategies have been explored, particularly within the framework of integrated multi-trophic aquaculture systems (Flannery 2017).

Several conceptual designs include:

- Biodegradable, constructions are designed to be fully sacrificial, completely submerged while the seaweed biomass remains attached.

The biomass is transported to deep-ocean sequestration sites via offshore advection on platforms made up of biodegradable subunits that separate and sink individually.

Seaweed can be released from self-anchored growing platforms using automatic or manual cutting methods, allowing it to sink on its own.

- Macroalgal biomass can be stored in geosynthetic containers for later deposition at depth, although this technique may be costly (Boettcher et al. 2019).

To the best of our knowledge, no nation has established a policy for reforesting the oceans within its exclusive economic zone. There are several ocean-related treaties that apply to international waters, including the Convention on Biological Diversity, the London Protocol, the International Seabed Authority, the International Maritime Organization, and the Law of the Sea. Although marine research in international

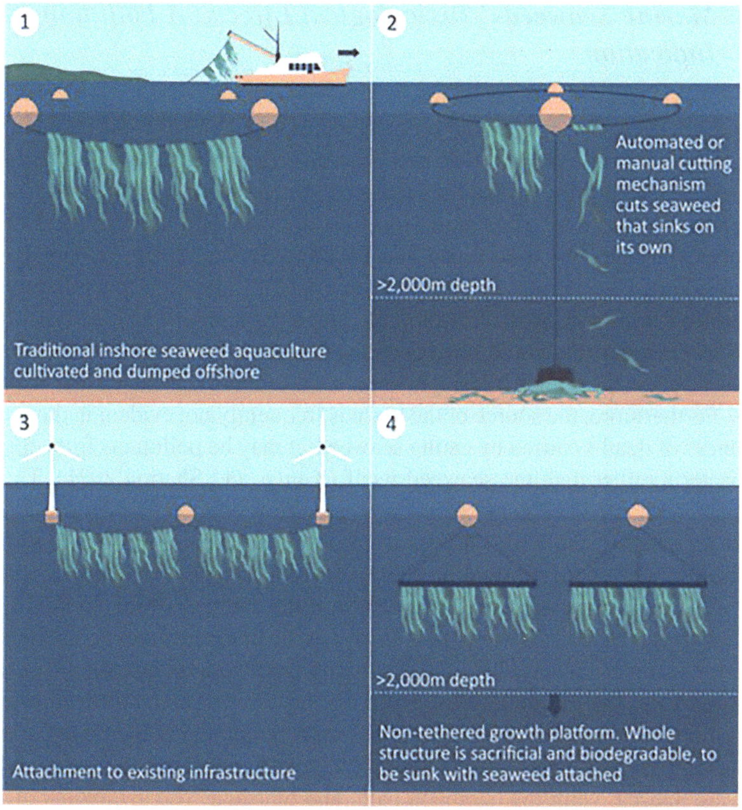

Fig. 3.1 Possible concepts for ocean afforestation. 1. Traditional nearshore aquaculture taken to be dumped offshore. 2. Self anchored growth platform sunk at the site. 3. Seaweed aquaculture infrastructure attached to existing offshore infrastructure (e.g. oil and gas infrastructure or a wind farm). 4. Free floating biodegradable platform (Rossi et al. 2022)

waters is covered by the Law of the Sea, there are few laws that directly address ocean afforestation. The main goal of the London Convention is to prevent pollution of the marine environment from entering the ocean. According to Article III b (ii) of the London Convention, "dumping" does not include the placing of material for a purpose other than simple disposal, as long as it does not go against the goals of the Convention. It is uncertain whether ocean afforestation falls under this category. Boettcher et al. (2019) believed that the socio-political hazards of ocean afforestation were moderate. According to the UN Convention on Biodiversity, the effects on biodiversity in international seas should also be taken into account, especially since carbon sink sites and seaweed growth sites may be in different locations.

3.1.2 Marine Seaweeds' Toxicological Effects: A Pollution Indication

Seaweeds: Seaweeds are important for food security and nutrition due to their essential biochemical content. However, they can also have toxicological effects, such as the release of harmful compounds that can affect marine life and human health. So when seaweed is used as a natural fertilizer or consumed as food, these harmful substances can enter the food chain and endanger human health (Dhargalkar and Verlecar 2009). Despite these risks, there is a lack of attention towards reports of seaweed toxicity, harmful species blooms, epiphytic bacteria and absorbed pollutants from seawater that could have detrimental interactions with medications and hormone levels in the human body, It is imperative to address the toxic and dangerous aspects. Furthermore, the source of the toxin is frequently not evident in the majority of instances of deaths caused by eating seaweed; it may be pollutants from epiphytic cyanobacteria rather than the seaweed itself (Kumar and Sharma 2021). The most commonly consumed and commercially valued seaweeds, such as *Porphyra* (nori), *Laminaria* (kombu), and *Undaria* (wakame), have not been linked to any poisonings. While heavy metal concentrations in edible seaweeds are generally below toxic levels (Cheney 2016). In this connection, Doty and Aguilar-Santos (1966) reported Caulerpicin from the green alga *Caulerpa racemosa* was the first poisonous component from seaweed. Customers reported symptoms that ranged from an anesthetizing feeling and numbness of the lips and tongue to, in one instance, fast and difficult breathing and numbness of the extremities. Also, the initial symptoms of the three *Gracilaria* food poisonings involving *G. verrucosa* and *G. chorda* in Japan in 1980, 1982, and 1993 were nausea, vomiting, stomach pain, and diarrhea. The symptoms of the *G. coronopifolia* poisoning in Hawaii in 1994 were again vomiting and diarrhea, but with the addition of a burning sensation of the mouth and throat (Nagai et al. 1996). However, citing three lines of evidence, Nagai et al. (1996) concluded that the true origin of the toxins was epiphytic cyanobacteria on the surface of the *G. coronopifolia* and not the seaweed itself. First, they discovered in microscopic observations that cyanobacteria were epiphytic on the surface of the sample of *G. coronopifolia* involved. Second, they were able to recover the two isolated toxins in a wash of the alga's surface. Third, both compounds had been previously isolated from marine cyanobacteria, including *Lyngbya majuscula* that had been reported from Hawaiian waters (Moore et al. 1984). As the consumption of processed seaweeds has grown, so too have concerns over their safety, in particular the concentration of two heavy metals, arsenic and iodine, which seaweeds have a unique ability to concentrate heavy metals in general, and arsenic and iodine in particular (Cheney et al. 2014). Almela et al. (2006), examined the arsenic, lead, and cadmium levels in 112 samples of 20 species of seaweed plus several seaweed preparations commonly used in soups and salads. Of the 112 samples, those of *Sargassum fusiforme* stood out because of their particularly high levels of total arsenic and inorganic arsenic (up to 149 mg/kg DW and 117 mg/kg DW, respectively. Amster et al. (2007) described how a 54-year-old woman from California developed symptoms of arsenic poisoning after taking two to four tablets of kelp (*Laminaria*) supplement per day for several months. Her symptoms included short- and long-term memory loss, fatigue, gastrointestinal disorders including diarrhea, nausea, and vomiting, and a rash on her lower legs.

Seaweeds are considered bioindicators of marine pollution and are one of the proposed biological quality elements for assessing ecological status. This is due to their high survival capabilities, ability to quickly react to changes in their environment, significant biomass, simple identification, long lifespan, and widespread availability (Garcia et al. 2020; El Zokm et al. 2022). Additionally, they are cost-effective and sensitive bioindicators for environmental monitoring and accumulate contaminants into their tissues directly from their surroundings and reflect the toxicological stress of their environment through measurable physiological, biochemical, and morphological changes. These toxicological responses not only affect the health and productivity of seaweed populations but also provide valuable insight into the ecological quality of marine environments so understanding their toxicological responses can help assess environmental health and inform mitigation strategies. Seaweeds, particularly species like *Ulva*, *Enteromorpha*, Fucus, *Gracilaria*, and *Sargassum*, are used globally to monitor environmental pollutants. According to Lindenmayer et al. (2023), seaweeds are valuablel biomonitoring tools for detecting trace metal pollution in Mediterranean marine environments. Seaweeds serve as biomarkers for heavy metals and micropollutants such as pesticides and hydrocarbon in the ecosystem (Ismail and Ismail, 2017; El Zokm et al. 2022). In particular, *Ulva* species are known for their ability to survive in both low and high-pollution areas year-round, making them suitable as medium-term environmental pollution monitors (El Zokm et al. 2022). Rakib et al. (2021) suggested that certain species may be more effective biomonitor for specific elements. For example, the brown alga *Sargassum* for As, Pb, and Mn monitoring, the red alga *Gelidium* for Cr, and the brown *Padina* for As, and Cu. El Zokm et al. (2022) found that Cladophora pellucida is an ideal biomarker for d-HCH contamination while U. compressa is a good indicator for a-HCH and methoxychlor. Ulva rigida and Gracilaria cornea have also been identified as effective biomonitors for capturing polychlorinated biphenyls (PCBs) and chlorinated pesticides (Pavoni et al. 2003). Furthermore, red seaweeds like *Corallina elongate* and *Gracilaria cornea*, as well as green seaweeds like U. rigida and Valonia utricularis, have been used to monitor organochlorine pesticides (OCP) (Vega-Moreno et al. 2007).

3.1.3 Wastewater Treatment

The sixth Sustainable Development Goal (SDG) focuses on water treatment and access to clean water. Marine ecosystems are impacted by the influx of organic pollutants such as pesticides, hydrocarbons, dyes, oils, greases, detergents as well as inorganic pollutants include heavy metals and nutrients. These contaminants originate from both man-made and natural sources (El Zokm et al. 2020; Ismail and El Zokm 2025). Numerous wastewater treatment technologies are being developed to address this issue. Despite the advancement of various technologies, seaweeds' adsorption technology has a significant impact on various industrial sectors due to their numerous advantages, including low cost, easy availability, large biomass, easy handling, high efficiency; dehydrated seaweeds could be stored for a longer duration without losing their properties, and inexpensive investment compared to other artificial techniques. Additionally, they are considered valuable bioindicators of environmental contamination due to their ability to absorb pollutants into their tissues. This enables them to accumulate a variety of pollutants from the ecosystem (Ismail and Ismail 2017; Saldarriaga-Hernandez et al. 2020). Sulfated polysaccharides are the primary reason for the strong ability of seaweeds to bind contaminants, such as trace metals. These polysaccharides are located in the fibril matrix and intercellular gaps of the algae. The hydroxyl, sulphate, and carboxyl groups of polysaccharides act as powerful ion exchangers, making them crucial sites for complexing metal cations (Arumugam et al. 2018). Also, algae also adsorb pollutants through the use of phytochelatins, which are peptides that are capable of chelating heavy metals (Inouhe 2005). According to Dokulil (2003), the pollutant adsorption process occurs in two steps. In the first step, the pollutant accumulates, by passive adsorption, on the outer surface of the seaweed thallus. In the second step, after being gradually absorbed through metabolic processes, these compounds/substances are ultimately converted into valuable commodities (Farghali et al. 2023). The main difference in the cell wall matrix and the significant differences in the chemical makeup of the cell walls across various seaweed species explain the variation in the affinity of different seaweed species for metal biosorption (Neveux et al. 2018). Furthermore, because old, decomposing fronds can contain significantly higher levels of pollutants, the quantity of pollutants in a given algae species may vary significantly depending on the stage of growth and/or the condition of the fronds (Maroli et al. 1993).

Actually, very few studies have utilized macroalgae to remove chemical oxygen demand (COD), biological oxygen demand (BOD), carbon fixation, lipid synthesis, total organic carbon (TOC), and turbidity from wastewater. This is primarily because most research focuses on removing colors, phenols, and heavy metals (Arumugam et al. 2018). In previous decay, seaweed was used as a biofilter in effluents of fish farms before being incorporated in fish diets, whereas they could be used in an integrated aquaculture system for both the treatment of wastes and the food supply for marine animals produced by aquaculture at the same site. Seaweeds possessed a great capacity to absorb nitrogen wastes (1.6 g of dissolved nitrogen/m^2, 0.7 g of NH_4 m^{-2}, 1.1 g of nitrates m^{-2}) (Deviller, 2003). Morais et al. (2020) utilized green *Ulva* and red *Gracilaria* seaweeds to diversify feed sources, reduce pollution, expand market resources, and lower nutrient levels in seawater effluents in order to increase profits. Ulvaceae green seaweed species accumulate more pollutants than other species, indicating that their unique accumulating activity is determined by their morphology, including branching, size, and blade type (El Zokm et al, 2022). *Ulva* and *Enteromorpha* proved the capacity to take in dissolved nitrogen, ammonium, nitrates, and nitrites contained in the waters of fish farms. These techniques, called the high-rate algal pond (HRAP), which improves the water quality, seems to decrease the mortality rate of the fish population submitted to farming over a long period. Neori et al. (2004) reported that *G. lichenoides* was used successfully for the treatment of wastewater from shrimp and fish farming, and the production of seaweed biomass by HRAP, which could be useful for the farming of species with a high economic value, such as abalone and urchins. The growth of Gracilaria lemaneiformis has the ability to absorb 128.10 tonnes of nitrogen, 15.89 tonnes of phosphorus, and 1192.03 tonnes of carbon from the seawater in Yantian Bay, per research by Duan et al. (2019). The contribution of seaweed farms to reducing nitrogen removal and coastal water eutrophication in China was calculated by Xiao et al. (2017). They discovered that growing seaweed in China removed 75 megatonnes of nitrogen and almost 9.5 megatonnes of phosphorus. According to the authors' projections, all phosphorus inputs into Chinese coastal waterways will have been eradicated by 2026 due to the seaweed industry. By 2050, World Bank estimates suggest that 500 million tonnes of seaweed will be harvested worldwide, requiring around 10 million tonnes of water nitrogen. This would account for 15 million tonnes of phosphorus, or around 33% of the phosphorus produced by fertilizers and manure, and 30% of the nitrogen that enters the ocean (Jagtap and Meena 2022).

Seaweed adsorption technology is also utilized for sewage treatment management and they also used in wastewater treatment as an adsorbent in order to replace the functional activated carbon (Das et al. 2025). They mostly used in wastewater treatment also reduces or removes toxic heavy metal contents as shown in Table 3.1 (He et al. 2008). Seaweeds also exhibit donor–acceptor contacts and a significant ability to adsorb phenolic chemicals via hydrophobic mechanisms (Antizar-Ladislao and Galil 2004). Navarro et al. (2008) examined the sorption of phenolic compounds by *Macrocystis integrifolia* and *Lessonia nigrescens* seaweeds. Both an electrostatic sorption process and a totally polar sorption pathway contribute to these species' greatest sorption efficacy of 35% at pH 10, according to the data. Adsorption of phenolic compounds onto the seaweed surface is facilitated by hydrogen bonding between the hydroxyl groups of the seaweed's polysaccharides, such as alginates, as noted by Garcia et al. (2020). The majorities of dyes used in the textile industry are resistant to aerobic digestion and do not biodegrade. *Sargassum wightii*-mediated dye degradation resulted in reduced water toxicity level, which is ensured by seed germination when the malachite green dye-treated water was used using *S. wightii* (Kaliaperumal et al. 2025). The brown algae *Fucus vesiculosus* is considered a suitable bio-adsorbent for efficiency removing organic dyes. This environmentally friendly and sustainable method of water remediation is due to its high bioadsorbent capacity to remove cationic dyes from aqueous solutions. It has been shown to achieve a removal rate of 98.71% for methylene blue (MB) and 96.68% for rhodamine B (RB) (Yadav et al. 2022). The green seaweed *Codium decorticatum* was utilized for crystal violet and Congo red dye from aqueous solution with high removal efficiency of 96.9% and 89.8%, respectively (Oualid et al. 2020).

The green seaweed *Codium decorticatum* was used to remove crystal violet and Congo red dye from aqueous solutions with high efficiency, achieving a removal rate of 96.9% and 89.8%, respectively (Oualid et al. 2020). Research demonstrating that seaweed species are efficient biomonitors has led to a surge in interest in the bioaccumulation of pesticide residues in seaweed in recent decades. *U. compressa*, a green

Table 3.1 Waste water treatment containing different pollutants using various seaweed species (Arumugam et al. 2018)

Seaweed	Type of wastewater	Pollutants		Efficiency
Sargassum sp.	Synthetic wastewater	Cd^{2+} and Zn^{2+}	Heavy metals	Cd^{2+} = 95.3% (acid treated) Zn^{2+} = 90.3% (acid treated)
Ulva rigida	Simulated wastewater	As^{3+}, As^{5+}, Sb^{3+}, Se^{4+} and Se^{6}		Se^{4+} = 0.5 mg/g (pH = 2–4) Se^{6+} = 0.2 mg/g (pH = 2–3) In raw forms, showed limited perspectives for arsenic removal but can be promising for selenium and especially antimony
Gracilaria sp.	Landfill leachate	As, Fe, Ni, and Cd		Fe = 100% Cd = 100% As = 100% Ni = 98%
Sargassum crassifolium *Gracilaria corticate* *Ulva lactuca*	Aqueous solution	Malachite Green	Dyes	95.7% (T = 25 C) 95.6% (pH = 8.0) 98% (t = 150min) 94.5% (T = 25 C) 93.8% (pH = 8.0) 97% (t = 150min) 93.3% (T = 25 C) 92.5% (pH = 8.0) 96% (t = 150min)
Caulerpa racemosa, Nizamuddin zanardini	Aqueous solution	Acid Black1(AB1) dye (Amino acid staining diazodye)		98.12% (Biomass = 5 g L 77.18 mg g^{-1}, pH = 2), 35.59% (Biomass = 5g L^{-1}, dye = 10 mg L^{-1}) 99.27% (Biomass = 5g L^{-1}, pH = 2)
Gracilaria parvispora, Nizamuddinia zanardinii	Pure textile methylene blue solution	Methylene blue		77.18 mg g^{-1} 565.96 mg g^{-1}
Gracilaria tikvahiae	Shrimp wastewater	Nitrogen	Nutrients	N = 35%
Ulva lactuca	Reject water from anaerobically digested sewage sludge	Nitrogen and Phosphorus		N = 22.7mgN g $DW^{-1}d^{-1}$ P = 2.7mg P g $DW^{-1}d^{-1}$
Chondrus crispus	Fin fishculture effluent	Nitrogen		Net N = 2.0 kgm^2 (at T = 6 and 13 °C)
Palmaria palmata				Net N = 2.0 kgm^2 (at T = 6 °C) Net N = 4.0 kgm^2 (at T = 16 °C)

algae, has demonstrated a noteworthy ability to accumulate OCPs and polycyclic aromatic hydrocarbons (PAHs) (Banach et al. 2020; El Zokm et al. 2022).

In general, the quantity of the evaluated micropollutants in the collected seaweed is correlated with the species, shape, concentration, and type of pollutant (El Zokm et al. 2022). Approximately 30–80% of the toxic chemicals present in wastewater, including caffeine, carbamazepine, ibuprofen and tris (2-chloroethyl) phosphate were biodegraded by seaweed cells (Ding et al. 2017). Seaweeds have sorbent qualities, making them a practical way to reduce the harmful effects of pollutants in aquatic environments, which is beneficial for the production of combined energy.

3.2 Economical Applications

3.2.1 Pharmaceutical, Medical and Cosmetic Applications

3.2.1.1 Pharmaceutical and Medical Applications

Pharmaceutical companies have begun to explore marine organisms in recent years. Marine-derived natural compounds play a major role in treating several human-related diseases due to their unique chemical diversity, biocompatibility, and excellent medicinal properties (Jeevithan et al. 2025). Seaweeds have long been known to possess medicinal activities since ancient times. Since the sixteenth century, China has utilized *Sargassum* to treat goitre, *Gelidium* to treat intestinal disorders, and *Laminaria* to assist in dilating the cervix during difficult deliveries. Bull kelp steam extract was utilized to treat headaches in Alaska, USA, while, *Durvillaea* was used to treat scabies in New Zealand, and dulse (*Palmaria palmata*) extract was used in eighteenth century England to help reduce fevers (Chapman and Chapman 1980). Between 1985 and 2008, over 1200 novel compounds were isolated from various classes of seaweed. Approximately 100–200 came from green species, 400–500 from brown, and up to 700 from red seaweeds. The majority of these compounds have shown promising bioactivities, indicating their potential as drugs (Shanmuganathan et al. 2015; Thennarasan et al. 2016; Ismail et al. 2023a).

The biological activities of macroalgae are related to their phytochemical contents which vary with species, and geographical sampling region due to changes in ecological parameters such as irradiance and nutrients (Punampalam et al. 2018). Red and brown species are the most interesting phyla pertaining to pharmacologically active compounds and widely used in the last decade (Ismail et al. 2016, 2023a).

There are numerous reports of unique compounds derived from seaweed with a broad range of bioctivities, like antioxidant (Gómez-Guzmán et al. 2018; Ismail et al. 2023a,b; El Newehy et al. 2025), antibacterial, antifungal (Barakat et al. 2022; Shobier et al. 2023; El-Said et al. 2024), antiviral (Shi et al. 2017; Ismail et al. 2022), antitumor (Ismail and Kanaan 2022; El Newehy et al. 2025), antidiabetic (Gómez-Guzmán et al. 2018; Ismail et al. 2024), anti-inflammatory (Vaughan et al. 2013;

Fig. 3.2 Scheme showing common medical applications of seaweeds (Own elaboration)

Shobier et al. 2023; Ismail et al. 2023a, 2024), Acetylcholinesterase inhibitory (Shanmuganathan et al. 2015), anticoagulant (Ismail and Amer 2020; Ismail and Kanaan 2022), antiobesity (Bedoux et al. 2014), Antimalaria or Antiplasmodial activity (Smyrniotopoulos et al. 2017; Elmaidomy et al. 2022), analgesic and antipyretic effects (Germain et al. 2011; Thennarasan et al. 2016) as illustrated in Fig. 3.2.

Antioxidant Activity

Antioxidant substances have garnered significant interest from the health and pharmaceutical industries. These compounds effectively prevent or delay the adverse effects that free radicals can produce. Huang and Wang (2004) classified antioxidants into two groups: reactive breaking antioxidants and preventive antioxidants. Additionally, they identified two categories of algal antioxidants: (1) Water-soluble antioxidants, such as vitamins, phycobiliproteins, and polyphenols; (2) Fat-soluble antioxidants, such as carotenoids and vitamin E (α-tocopherol).

Many researchers are interested in finding natural antioxidants with safety and effectiveness, that can be substitute the synthetic antioxidants such as butylated hydroxyl anisole (BHA), butylated hydroxyl toluene (BHT), α-tocopherol and propyl gallate, which are responsible for several side effects such as liver damage and carcinogenesis (Ismail et al. 2023b; El Newehy et al. 2025). Carotenoids, polysaccharides, vitamins, phenol, fatty acids, and phycobiliproteins from macroalgae have shown great antioxidant activity (Ismail et al. 2016; 2023a,b,c).

Phenolic compounds are widely distributed in all seaweed species and act as antioxidants that stop lipid peroxidation, scavenge Reactive Oxygen Species (ROS), and chelate metal ions to capture the lipid alkoxyl radical (Gómez-Guzmán et al. 2018). The intensity of the antioxidant activity of these complex polyphenols is related to the degree of polymerization of the polyphenol. Lower degrees of polymerization result in greater antioxidant activities. Phlorotannins have shown antioxidant ability by enhancing antioxidant defenses and/or scavenging ROS (Ismail et al. 2023a). The antioxidant capacity of macroalgal phenolic compounds helps to mitigate the harmful effects of advanced glycation end products (AGEs). AGEs are toxic byproducts associated with hyperglycemia in diabetes, causing damage to a patient's tissues and organs (Gómez-Guzmán et al. 2018).

The polysaccharides from three different classes have broad-spectrum therapeutic properties. Specifically, fucans, carrageenans, and ulvans have been shown to exhibit strong antioxidant and immunostimulatory effects (Devaki et al. 2009; Barakat et al. 2022; Ismail et al. 2024, 2023a; El Newehy et al. 2025). The antioxidant activity of these polysaccharides is influenced by their structural characteristics, including the molecular weight, degree of sulfation, type of the major sugar and branching. Sulfated polysaccharides with a low molecular weight, for instance, have shown to have higher antioxidant activity compared to those with a high molecular weight. This is because smaller polysaccharides can more easily enter cells and efficiently provide protons (Ismail et al. 2022). Devaki et al. (2009) observed an increase in the antioxidant potency of rats when given *Ulva lactuca* polysaccharide extract orally at a dose of 200 mg/Kg body weight daily for 21 days.

Antimicrobial Activity

Seaweeds exhibit antimicrobial activity due to a variety of bioactive compounds they produce such as polysaccharides, fatty acids, phenolic compounds and halogenated compounds (Shobier et al. 2023; El-Said et al. 2024). Their antimicrobial action might be due their ability to alteration of microbial cell permeability that led to the loss of internal macromolecules or by the interference with the membrane function and loss of cellular integrity and eventual cell death (Abu-Ghannam and Rajauria 2013).

Based on taxonomic groups, Al-Saif et al. (2014) indicated that the extracts of red alga *Gracilaria dendroides* from Jeddah coast of Red sea, Saudi Arabia was more efficient against pathogenic human bacterial strains followed by green alga *Ulva reticulata*, and brown algae *Dictyota ciliolata*. In contrast, Omar et al. (2012) who detected that Phaeophyta spp. from Jeddah coast of Red Sea, Saudi Arabia had the highly antibacterial active against the tested bacteria followed by Chlorophyta spp. and Rhodophyta spp.

Numerous scientific findings indicate that the antibacterial potential peaked in the spring, most likely due to the presence of active chemicals during that time of year. There was a noticeable variance in the phenolic content, antibacterial activity, and antioxidant activity, with the spring and summer showing the greatest inhibitory effects (Pérez et al. 2016). Sulfated Polysaccharides from *Corallina officinalis* and *Pterocladia capillacea* showed antibacterial activity toward gram-positive bacteria (Ismail and Amer 2020). Phycobiliproteins from *C. officinalies* also demonstrated antibacterial action against various human pathogenic bacteria, contingent on the bacterial type and dosage (Ismail et al. 2023c).

Padmakumar and Ayyakkannu (1997) screened 80 species for human infections. They discovered that only 27.5% of the investigated species had antifungal activity, while roughly 70% had antibacterial activity. The various extracts of *Cy. crinita* and *Ulva intestinalis* showed significant antifungal efficacy toward *Candida krusei* (Berber et al. 2015). Pourakbar et al. (2021) observed the antifungal activity of *Gracilariopsis persica* toward various plant pathogenic fungi. The ethanolic extract of *Ulva linza* exhibited the highest antifungal activity against *Candida albicans* (Shobier et al. 2023).

Antiviral Activity

Seaweeds show potential as natural sources of antiviral chemicals, which could pave the way for the development of new antiviral treatments in vitro and in vivo (Dinakarkumar et al. 2025). Metabolites from seaweed can act as antiviral agents by either enhancing the immune system of the host cell or inhibiting the virus from replicating before it can infect the host cells (Lomartire and Goncalves 2022). Compared to ribavirin, a medication used to treat the Chikungunya virus, red alga *Osmundaria obtusiloba* has been shown to have stronger antiviral and virucidal effects (Cirne-Santos et al. 2019).

Shi et al. (2017) illustrated the antiviral activity of macroalgal polysaccharides towards viral infection by targeting the viral attachment stage usually using the following two strategies: interacting directly with virions or mimicking the virus-associated protein (VAP) to bind to the respective receptors as explained in Fig. 3.3. Other antiviral machenisms of macroalgal polysaccharides were found to be immunomodulators that activate Natural killer cells (NK) cells or induce antiviral immune responses, which helps the inhibition of viral infection and acceleration of viral clearance (Shi et al. 2015).

For example, fucoidans from brown species stimulated many non-specific and specific immune responses under in vitro and in vivo conditions. Also, fucoidans can not only improve the activation of NK cells and enhance the Th1-type immune response, production of antigen-specific antibodies, and generation of memory T cells (Shi et al. 2015). Sulfated polysaccharides derived from *S. euryphyllum* and brown *S. aquifolium* have been suggested as natural resources to fight the avian infectious bronchitis virus (IBV). These sulfated polysaccharides prevent the virus

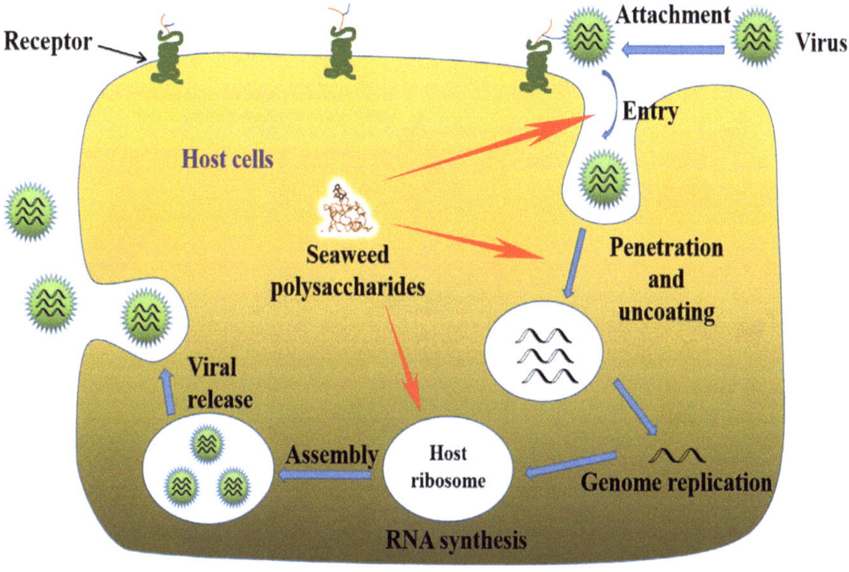

Fig. 3.3 The process of viral infection and the antiviral phase of macroalgal polysaccharides (Shi et al. 2017)

from entering cells, most likely by preventing its attachment to host cell receptors (Ismail et al. 2022).

Phlorotannin derivatives obtained from the brown alga *Ecklonia cava* demonstrated inhibition of protease and reverse transcriptase (RT) enzyme, which play a vital role in virus replication inside the host cells. When obtained from various sources, they may elicit different inhibitory properties to these target enzymes (Singh and Kumar 2022).

Griffithsin (GRFT) is a protein that was first isolated from *Griffithsia* sp. It has shown potential as a good candidate for the development of topical antiviral agents. Additionally, GRFT inhibits HCV, is and has been found to be effective against enveloped viruses such as the Japanese Encephalitis Virus (JEV), HSV-2, and HPV (Singh and Kumar 2022).

Anticancer and Antiproliferative Activities

The antiproliferative effect of algal polysaccharide fractions may be influenced by their antioxidant activity (Ismail and Kanaan 2022; El Newehy et al. 2025). Furthermore, fucoidans activate the host immune system against tumors and inhibit the growth of HCT-15 human colon carcinoma cells through various apoptotic events such as DNA fragmentation, chromatin condensation, and an increase in the population of sub-G1 cells. They also mediate apoptosis signaling through the mitochondrial pathway, as explained in Fig. 3.4 (Hyun et al. 2009). Fucoidan from *Sargassum*

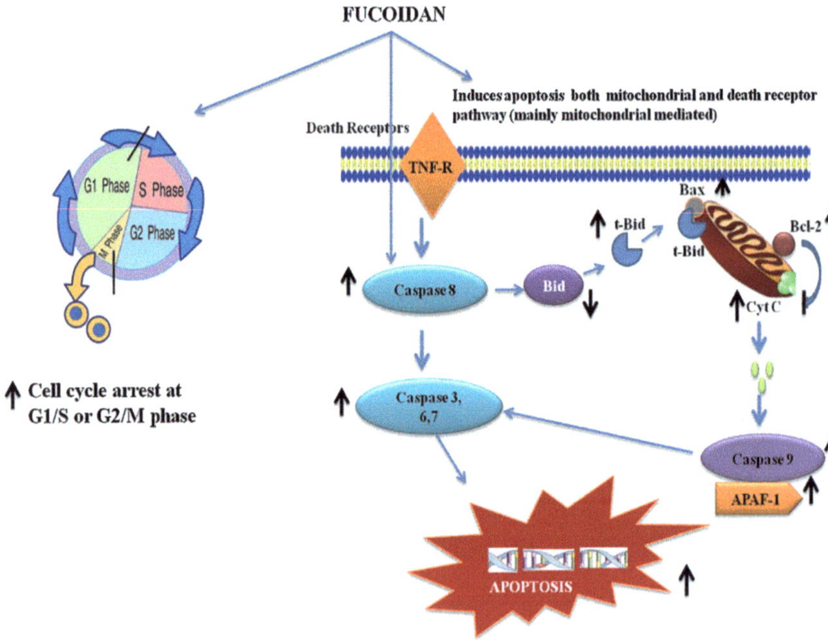

Fig. 3.4 Schematic representation of cell cycle arrest and apoptosis mechanisms of fucoidan. Fucoidan arrests the cell cycle at G1/S or G2/M phase. Apoptosis occurs via intrinsic and extrinsic pathways. Fucoidan can activate both pathways leading to apoptosis. Black small arrows indicate down regulation, while upper arrow indicates up regulation of protein activation by fucoidan (Hyun et al. 2009)

cinereum and *Turbinaria decurrens* and F-AuNPs reveal remarkable anticancer properties against HepG2, THP-1, and BNL cells (El Newehy et al. 2025). Furthermore, fucoidans from the brown algae *Dictyota dichotoma* reduced the development of cancer cell colonies at 200 μg/mL and showed no cytotoxicity below 400 μg/mL (Usoltseva et al. 2018). The aqueous extracts of *Cy. crinita*, *Caulerpa sedoides* and *Caulerpa compressa* showed significant antiproliferative activity against both human tumor cell lines HCT15 and MCF7 due to their phenol contents (Mhadhebi et al. 2014). Sulfated galactans exhibited the maximum antiproliferative potency against HeLa cervical cancer cell lines at 1.5% (Ismail and Kanaan, 2022). Another important metabolite, namely Kahalalide F (peptides), derived from green alga *Bryopsis* sp. and Caulerpenyne obtained from *Caulerpa* sp., showed anticancer and antiproliferative effects against human prostate, lung, and colon cancer (Scheuer et al. 2000). Macroalgal polyunsaturated fatty acids such as omega-3, eicosapentaenoic acid (EPA), and docosahexaenoic acid (DHA) have also demonstrated anticancer activity (Francavilla et al. 2013). α R-phycoerythrin from *Pyropia yezoensis* shows promise as a natural antioxidant and anticancer medication for food and pharmaceutical uses (Ulagesan et al. 2021).

Antidiabetic Activity

A malfunction in insulin action, secretion, or both can lead to diabetes mellitus, a metabolic disease. It is one of the most severe metabolic diseases in the world. It is usually classified into two main groups i.e. Type I diabetes mellitus (T1DM), caused due to complete absence of insulin production and Type II diabetes mellitus (T2DM), due to the relative deficiency of insulin secretion and tissue resistant to the insulin action1 (Cefalu 2001).

Macroalgae has potential as antidiabetic agents (Gómez-Guzmán et al. 2018; Ismail et al. 2023a, 2024). The weekly dietary consumption of *Porphyra yezoensis* and *U. pinnatifida* led to lower incidence of diabetes in Korean men also the risk for developing T2DMin Korean men may be reduced by dietary consumption of seaweed (Lee et al. 2010).

However, seaweeds have α-amylase and α-glucosidase inhibitory activities (Venkatraman et al. 2018; Ismail et al. 2022, 2024). Especially, the marine brown seaweed species are a rich source of extractable bioactive compounds such as pigments, fucoidans, and polyphenols, which have positive impacts on diabetes (Ismail et al. 2023a, 2024). *Sargassum oligocystum* improved diabetes by reducing the insulin resistance, decreasing glucose concentration, and regeneration of pancreatic damaged β-cells (Akbarzadeh et al. 2018). *Turbinaria conoides* had a potentially important role in managing diabetes by the inhibition of α-amylase and nonenzymatic glycosylation of haemoglobin (Venkatraman et al. 2018). *Padina tetrastromatica* showed good antidiabetic properties, such as increasing the insulin secretion and enhancing the insulin-stimulated glucose uptake, and modulating the carbohydrate metabolism (Ponnanikajamideen et al. 2017). Moreover, brown seaweeds *"P. sulcata, S. binderi* and *T. conoides*" displayed potent inhibitory effects towards Dipeptidyl peptidase 4 (DPP-4) and promoted GLP-1 secretion (Chin et al. 2014).

According to recent data, seaweed phenolic components may have antidiabetic effects because they can inhibit specific digestive enzymes. For example, phlorotannin components of A. nodosum may have antidiabetic activity by inhibiting both a-amylase and a-glucosidase, which reduces the absorption of carbohydrates, and PTP1B, which increases sensitivity of insulin (Gómez-Guzmán et al. 2018; Ismail et al. 2023a). Phlorotannins isolated from Phaeophyta species enhance the efficiency of antioxidant enzymes like catalase, glutathione peroxidase and superoxide dismutase and which can also account for its cell protective effects in high-glucose concentrations (Lee et al. 2013; Ismail et al. 2023a).

Sulfated polysaccharides had the ability to regulate the glucose homeostasis, improve insulin sensitivity, increase glucose uptake, and reduce blood glucose levels (Kim et al. 2021). Fucoidan from brown seaweeds decreased the rat blood sugar level (Yang et al. 2017). Laminaria hyperborea alginate lowers blood sugar, increases insulin, and suppresses hunger (Torres et al. 2020).

There are other bioactive compounds were been isolated from seaweed with antidiabitic properties, such as, benzene acetamide, bromophenols, n-hexadecanoic acid, 2-piperidione, and polysaccharide derivatives from red species (Polysiphonia urceolata, Rhodomela confervoides, and Symphyocladia latiuscula) exhibited hypoglycemic potentials by inhibiting α-glucosidase (Seung-Hong and You-Jin 2013). Aastaxanthin has a protection against diabetes through inhibition of starch-digestive enzymes (Lee et al. 2010).

Anti-inflammatory/Anti-arthritic Activity

Inflammation is a defensive response that occurs in a wide variety of physiological and pathological processes. It can be caused by stress, injury, or infection, or it may be triggered by the release of chemicals from tissues and migrating cells (Khursheed et al. 2023). The macroalgal bioactive components act as a competitive inhibitor of cyclooxygenase and/or lipoxygenase in an inflammation reaction, resulting in decreased production of prostaglandins and leukotrienes (Gunerken et al. 2015; Khursheed et al. 2023). The remarkable anti-inflammatory activity of *Sargassum* extract is achieved through a variety of mechanisms. Firsty, it has an indirect effect through antioxidant substances like sulfated polysaccharide and polyphenol. Secondary, it contains a high concentration of metals and minerals, including sulfur, halide compounds such as bromine, chlorine, and iodine which are unique compounds for marine habitat (Vaughan et al. 2013).

According to the World Health Organization (1989), Egyptian seaweeds are safe and important to prevent inflammatory, cardiovascular, and nervous system disorders (Ismail et al. 2017; Shobier et al. 2023). Also, a low-dose crude sulfated polysaccharide from *Sargassum ilicifolium* (5 mg/kg) is an efficient anti-arthritic agent that confers its effects via the cytokine pathway (Ramamoorthi et al. 2025).

The anti-inflammatory features of bioactive compounds like curcamin, resvatol, capsaicin, catechins, vitamins, β-carotine, tannins, flavonoids and polyphenols to their abilities to moderate cell Mitogin-Activated Protein Kinases (MAPK) signalling pathways, proliferation, apoptosis, redox balance or suppressing an effect of NF-kB and AP-1 transcription (Hatcher et al. 2008). Algal flavonoid compounds had anti-inflammatory effect through modulation of NFκB (Chao et al. 2013). Phlorotannins extracted from different brown species revealed potent anti-arthritis activity relating to the tested concentration (Ismail et al. 2023a).

Polysaccharide fractions of *Gracilaria verrucosa* were administered to mice orally and intraperitoneally and increased antioxidant activity as well as anti-inflammatory activity stimulating phagocytosis (Gunerken et al. 2015).

Analgesic and Antipyretic Activities

An analgesic "painkiller" agent that used to achieve analgesia. Algal terpenes, peptides and sulphated carbohydrates that exhibit analgesic effects (Paul and Muthu 2016). Thennarasan et al. (2016) recorded the analgesic activity of methanol extract of brown alga *Lobophora variegata* by reducing the number of acetic acid-induced writhes as the diclofenac and related drugs that can inhibit cyclooxygenase in peripheral tissues, which may be due to the release of endogenous substances that stimulate pain in nerve endings similar to paracetamol and other Nonsteroidal anti-inflammatory drugs (NSAIDs). The methanol extracts of *Sargassum swartzii* and *U. reticulata* commonly used as oral analgesic agent without side effect and had similar to impacts of aspirin (100 mg/Kg) and morphine (100 mg/Kg) (Hong et al. 2011).

Antipyretics are agents that reduce elevated body temperature. Bioactive compounds from macroalgae act as novel and safe antipyretic agents, such as flavonoids like baicalin which showed antipyretic impact by suppressing Tumor Necrosis Factor (TNF-α) and inhibiting arachidonic acid peroxidation. This results in reduced prostaglandin levels, blocking cyclooxygenase enzyme activity and/or inhibiting mediators of pyrexia similar to paracetamol. Ultimately, this reduces fever and pain (Germain et al. 2011; Chandrasekharan and Simmons, 2004). The methanol extract of *L. variegata*' showed significant antipyretic activity as paracetamol (Thennarasan et al. 2016). *U. rigida* has an inhibitory activity on prostaglandins, which set the thermoregulatory center lower temperature biosynthesis (Chandrashekaran et al. 2014).

Acetylcholinesterase Inhibitory

Reduced levels of the neurotransmitter acetylcholine (ACh) are a hallmark of Alzheimer's disease (AD), a degenerative and progressive condition that primarily affects the cerebellum and neocortex, which are important for mental processes.

This illness can cause cognitive problems, behavioral abnormalities, and memory loss. By 2040, there will be over 80 million people with AD if appropriate treatment is not received. Approximately 36.5 million people worldwide suffer from dementia, with two-thirds of those affected living in developing nations (Alzheimer's Association 2014).

The cholinergic hypothesis states that a major contributing factor to AD symptoms is a considerable decrease in the cholinergic neurotransmitter acetylcholine (ACh) in the central nervous system (Alzheimer's Association 2014). The main therapeutic strategy against AD is acetylcholinesterase inhibition. Only a few literatures have been reported regarding the AChE inhibitory effect of macroalgae.

The AChEI activity of plant extracts is classified according to Vinutha et al. (2007) as potent inhibitors (greater than 50% inhibition), moderate inhibitors (30–50% inhibition) and weak inhibitors (below 30% inhibition). According to this classification, the extracts of ***Ochtodes secundiramea*** exhibited moderate activity ($48.59 \pm 0.8\%$), while ***Hypnea*** *musciformis* and *P. capillacea* extracts have weak activity (7.21 and 5.38%, respectively), at a higher concentration which may be related solely to halogenated monoterpenes (Machado et al. 2015). The ethanol extract of most seaweeds acts as an inhibitor of amyloid-aggregation, which is in Alzheimer's disease (Shanmuganathan et al. 2015). *Rhodomelopsis africana, Halimeda cuneata,* and *Codium duthieae* potential neuroprotective activities of these extracts which could be linked to the presence of some phenolic compounds so these seaweeds act as a good source of antioxidants, cholinesterase and β-amyloid aggregation inhibitors and could be explored as an alternative therapeutic strategy for the management of AD (Olasehinde et al. 2019). Methanolic extract of *H. cuneata* exhibited the greatest bioactivity as AChE inhibitors ($IC_{50} = 70$ μg/mL) (Rengasamy et al. 2015).

Anticoagulant and Fibrinolytic Activity

Anticoagulants known as blood thinners, prevent the formation of blood clots and decrease the threat of stroke, cardiac failure, and obstruction in the veins and arteries. The anticoagulants and fibrinolytic ability of macroalgae may be related especially to their polysaccharide's composition, their molecular weight, sulfate content, type of glycosidic bonds, position, uronic acids monosaccharide composition, and the degree of polysaccharides chain branching. These polysaccharides carrying a negative charge which gives the ability for binding calcium ions and therefore prevents the formation of a clot (Kuznetsova et al. 2021) While, fibrinolysis is a major biological process that functions to prevent unwanted and undesirable fibrin clots from forming during blood circulation under normal physiological conditions (Fibrinolysis is a major biological process that functions to prevent unwanted and undesirable fibrin clots from forming during blood circulation under normal physiological conditions (Jeevithan et al. 2025).

Sulfated polysaccharides (SPs) have been recommended in numerous studies for their anticoagulant, antithrombotic, and fibrinolytic properties, which help prevent thrombosis and aid in thrombolysis (Wu et al. 2016; Ismail and Amer 2020; Kuznetsova et al. 2021). While direct thrombin inhibition and competitive binding of fibrinogen to prevent thrombin's activities are potentially effective, fucoidan's anticoagulant properties are typically mediated through the activation of thrombin inhibitors (Wu et al. 2016). Fucoidan from different brown species like *Ascophyllum nodosum*, *Fucus distichous*, *F. evanescens*, *F. serratustus*, *Laminaria digitata*, *Saccharina latissima* and *S. longissima* had strong anticoagulant activities in vitro and in vivo models depending on their content and/or positioning of sulfate groups, as well as the molecular weight of the polymers (Holtkamp et al. 2009). Also, the anticoagulant activity of sulfated polysaccharides from *Corallina officinalis* and *Pterocladia capillacea* exhibited more activity than heparin (Ismail and Amer 2020). There is commercial *Fucus vesiculosus* fucoidans (from the Sigma-Aldrich Chemical Co) are present in Market (Pereira 2018). Seaweed polysaccharides (PSs) have anticoagulant and antithrombotic properties that impact the extrinsic and intrinsic coagulation pathways as well as the final stage of coagulation, the conversion of fibrinogen to fibrin under the influence of thrombin (Kuznetsova et al. 2021). Green algae of the genus *Codium* and the brown alga *Sargassum fulvellum* contain metabolites and enzymes wit fibrinolytic activity due to the presence of protease activators (PAs) and substances that promote the breakdown of fibrin clots (Wu et al. 2016). The protease produced by *Codium latum* has a fibrinolytic activity of 314 mm^2, which is surprisingly higher than that of potent snake venoms and can hydrolyze a fibrin clot (Kuznetsova et al. 2021).

Anti-obesity Activity

Obesity is defined as the abnormal or excessive accumulation of fat that poses a risk to health and results in an increased risk of many serious diseases such as hyperlipidemia, hypertension, type II diabetes, and cardiovascular disease.

Actually, there are some macroalgal products toward anti-obesity such as, PolymannTM (Korea Bio Polymer Co. Ltd), is one product in that based on polymannuronic acid from the kelp *Undaria* that helps in controlling blood cholesterol (Hotchkiss 2011). LipoxanThin™ is a natural drug consisting of fucoxanthin produced by the National Bioscience Corporation (Chester, New York), which is an organization that has received several patents for the creation of natural products used in the treatment of obesity, fatty liver, and hyperlipidemia have the potency to counteract a metabolic syndrome X and improved insulin sensitivity and blood glucose (Holt 2008).

In general, The fiber content of seaweeds can be as high as 75% of the total algal dry weight which comprises the structural polysaccharides which have been found to produce hypocholesterolemic and hypolipidemic responses as they tend to reduce cholesterol absorption in the gut. *Sargassum polycystum* powder treatment possessed a positive impact on the inhibition of weight gain and posed promising value in preventing obesity due to the high soluble dietary fiber content and its high amount of fiber ($16.66 \pm 0.18\%$) and low lipid content ($3.71 \pm 0.36\%$) (Matanjun et al. 2010). Bedoux et al. (2014) detected the minerals, proteins and carbohydrates from *Laminaria digitata* were used in slimming and anticellulite formulations. Fucoidan from *Fucus vesiculosus* and *Pelvetia wrightii* are frequently used in slimming and anticellulite formulations (Bedoux et al. 2014). 0.02% Fucoxanthin significantly lowered body weight, body fat accumulation, visceral fat-pad weights, white adipose tissue weight gain, the size of adipocytes in diabetic mice and increased brown adipose tissue weight in mice. Fucoxanthin also reduced the activities of adipocyte fatty acid synthesis, hepatic fatty acid, and triglyceride production, as well as plasma and hepatic triglyceride levels. Both fucoxanthin and fucoxanthinol prevented lymphatic triglyceride absorption and the increase in bloodstream triglyceride levels, likely due to their inhibition of gastrointestinal tract lipase activity (Matanjun et al. 2010; Park et al. 2011) or relating to oxidation of fatty acids, heat production, and energy dissipation through upregulating the expression of uncoupling protein in the white adipose tissue (Maeda et al. 2009). Also, fucoxanthin promoted mRNA expression of β3-adrenergic receptor in a white adipose tissue of obese mice, which was responsible for a lipolysis and thermogenesis (Miyashita et al. 2011).

Antimalaria or Antiplasmodial Activity

Malaria is one of the most important infectious diseases worldwide. The antimalarial properties of macroalgae have only been somewhat documented. A prospective lead molecule for future antimalarial drug development is the green algae *Halimeda*, from which Elmaidomy et al. (2022) isolated two novel compounds and four previously described ones The antiplasmodial activity of ethanolic extracts from different algal species against *Plasmodium falciparum* might be due to the presence of polysaccharides, proteins, and phenols (Ravikumar et al. 2011). The Irish brown algae *Bifurcaria bifurcata* producesa novel linear diterpene called bifurcatriol.This compound has been found to suppress drug-resistant P. falciparum K1 at low concentrations (0.65 μg/mL) (Smyrniotopoulos et al. 2017). The mechanism of action may be due to the inhibition of *P. falciparum* merzoite invasion into the erythrocytes or disruption of *P. falciparum* rosettes (Adams et al. 2005).

Macroalgae Thyroid Supplements

Iodine is a vital nutrient required for proper growth for all age groups and is essential for the normal regulation of thyroid function, which also involves the brain and pituitary gland. Furthermore, iodine is extremely crucial in pregnancy and breast feeding for the normal development of the brain cells of the baby and can lead to weak intellectual growth and abnormal development of the brain (Zimmermann 2012). A deficiency of iodine in the body can result in abnormalities such as thyroid enlargement or goiter, hypothyroidism, and mental retardation (Di Liegro 2008).

Seaweeds with their high iodine content could be useful to remedy dietary iodine deficiency without the side effect of hypertension brought on by salt, which it absorbs in vast amounts from the seawater. The iodine content in brown seaweeds is higher than as compared to red and green seaweeds (El Zokm et al. 2021). The consumption of excess seaweed may increase the quantity of iodine above acceptable levels. This can raise the levels of thyroid-stimulating hormone, which can cause serious conditions like thyroid and goiter and can heighten the risk of thyroid cancer (Zava and Zava 2011).

Seaweed as Contraceptive Material

Different types of seaweeds have shown varying contraceptive effects. In female rats, red algae *Gracilaria corticata* and *Gelidiella acerosa* from Sri Lanka's coastal waters have been proven to be safe and effective post-copulatory contraceptives. Furthermore, sperm motility was totally suppressed by ethanolic extracts of *Gracilaria edulis*, suggesting the presence of a spermicidal chemical that breaks down sperm plasma membranes (Dolui 2021). Human sperm was completely inhibited after 20 s of contact with *Halimeda gracilis* at a dosage of 10 mg/ml. this is due to secondary metabolites like sugars, flavonoids, alkaloids, and proteins cause damage to the plasma membrane (Prakash et al. 2014).

Other Medical Applications

Macroalgae consumption led to decreasing blood pressure, may be its linked the hypotensive effects to their dietary fiber or their rich nitrate content (Bocanegra et al. 2009). Phlorotannins from brown seaweeds have signficant effect on hypertension due to their vasodilator effects and angiotensin I converting enzyme (ACE) inhibitory activity (Park et al. 2017). Fitzgerald et al. (2011) ascribed the antihypertensive effects to the bioactive peptides in seaweeds. In addition to improving blood vessel circulation and preventing deadly diseases such as peripheral arterial disease, atherosclerosis, and heart failure, seaweeds also help keep the heart healthy. Seaweed secondary metabolites act as hypoglycemic effects, lowering blood pressure, controlling cholesterol levels, limiting the production of hepatic cholesterol, preventing hyperplasia in the gastrointestinal tract and promoting regeneration. Fucoxanthin protects stroke-prone spontaneously hypertensive rats with cerebrovascular diseases from ischemic neuronal cell death (Peng et al. 2011).

Polyunsaturated fatty acid from seaweeds, particularly, eicosapentaenoic acid, arachidonic acid and docosahexaenoic acid have a dynamic role in human health including cardiovascular effects, coronary heart diseases, antiarteriosclerosis, antihypertensive, antiarrhythmic responses, as well as regulation of membrane fluidity, electron and oxygen transport (Peng et al. 2011). Additionally, macroalgal vitamin C helps prevent of cardiovascular illnesses, boosts the immune system, promotes the absorption of iron in the intestines, and traps free vitamin E to prevent low-density lipoprotein oxidation (Bender, 2003; Ismail et al. 2025).

3.2.1.2 Cosmetic Applications

Seaweeds may be a desirable and cost-effective source for the cosmeceutical industry (Zárate et al. 2020). Due to their special chemical properties, bioactive constituents can be utilized as natural cosmetics or added to cosmetic formulas as organic dyes, texturing stabilizers or emulsifiers. These materials of biomolecules relevant to skincare (Pimentel et al. 2018), and they have the potential to have minimal cytotoxic effects on humans (Thiyagarasaiyar et al. 2020).

Seaweeds are used as photoprotective ingredients in sunscreen because they typically produce secondary metabolites that shield their cells and organelles from UV radiation. Moreover, some studies have detailed the many advantages of seaweed compounds as low-cytotoxic topical skin care solutions (Kageyama and Waditee-Sirisattha 2019). Cosmeceuticals containing seaweeds are non-irritating and can perform various functions. They can improve conditions such as psoriasis and prevent skin problems, like atopic dermatitis, hyperpigmentation, acne, wrinkles, and hair loss (Huang et al. 2019; Park 2020). Additionally, seaweeds were utilized in preparations for skin-whitening, anti-aging, and anti-pigmentation (Polat et al. 2021).Seaweeds including *Fucus*, *Laminaria*, and *Chondrus* are very benefical for replenishing and nourishing the skin (Jesumani et al. 2019). These genera are the most commonly used algae genera in cosmetics (Pagels et al. 2022). A cosmetic formulation containing 5–10% extract of *Laminaria japonica* improved skin moisture (Pimentel et al. 2018). In a study of 23 volunteers, Park et al. (2021) found that extracts from the red seaweed Pyropia yezoensis were effective and safe for skin whitening, wrinkle prevention, decreasing tyrosinase activity, stimulating collagen synthesis, and improving skin brightness. Seaweed components can also replace various additives, such as antimicrobials or conventional ultraviolet ray blockers (Han et al. 2017). In cosmeceuticals and pharmaceuticals, alginates are widely used as a variety of gelling agents, such as protective colloids, emulsion stabilizers, ointment bases, lotions, pomades, hand jellies, hair products, facial creams, and beauty masks (Dolui 2021; López-Hortas et al. 2022). More than 20% of products containing carrageenan are used in pharmacies and cosmetics for personal grooming and hygiene. These products include toothpaste, hair conditioners, medications, lotions, deodorants, shampoos, foams, shaving creams, sunscreen, and sprays (Morais et al. 2020). The brown alga Laminaria produces laminaran, a polysaccharide mainly used as an anti-cellulite product due to its numerous bioactive qualities (Morais et al. 2021). Fucoidans found in brown algae may also be utilized as an anti-aging medication. This is because they enhance heparin-growth factor production, increasing cell hydration and flexibility while promoting tissue and cell growth (Dolui 2021).

Several scientific studies have demonstrated the potential skincare benefits of seaweed bioactives, which is why they have recently been added to cosmetic products. In addition to deep cleaning, seaweed-based cosmetics can tighten tone, increase blood and lymph circulation, and enhance the suppleness and moisture content of the skin. Their antioxidant and nutritional qualities aid in the skin's absorption and retention of minerals (López-Hortas et al. 2022). The incorporation of seaweed components has been successful in various physical forms and is commercially available in soaps, shampoos, sprays, hydrogels, and creams (López-Hortas et al. 2022). Additionally, seaweed extracts can be combined with extracts from terrestrial plants, medicinal herbs, mushrooms, microalgae, and fish, as well as with conventional ingredients or even gold (Park, 2020). These mixtures of species and combinations with other raw materials during the manufacturing of cosmetics can be adopted to generate synergistic effects (Bi et al. 2018; López-Hortas et al. 2022).

Hydrochloidal substances from red and brown seaweeds are commonly used as thickeners in many cosmetics and hair products as explained in Table 3.2. Polysaccharides have an important role in cosmetic formulations as humectants and moisturizers and can be linked to keratin through hydrogen bonds, thus improving skin moisturization. Polysaccharides of *Pyropia yezoensis* were proven to protect against UVA-induced photoaging, and its peptide was found to promote collagen synthesis in human skin fibroblasts (Kim et al. 2017). Fucoidan can also produce a protective layer on the skin, preventing water evaporation (Yu and Gu 2015). Choi et al. (2013) detected that algal alginate and protein attach to skin proteins to form a protective barrier for moisture loss regulation. *Saccharina japonica* polysaccharides had better moisturizing properties and could be an interesting ingredient for cosmetics (Wang et al. 2013c). *Plamaria* and *Prophyra* contain arginine that is a precursor of urea, which is a component of Natural Moisturizing Factor (NMF), used in cosmetic formulations (Bedoux et al. 2014). Some of the Australian companies like Maritech and Marinova have stated the clinical usefulness of their macroalgae extracts in reducing wrinkles and improving skin appearance of aging skin (Bedoux et al. 2014).

Macroalgae has emerged with mechanisms to respond against the hazardous effects of ultraviolet 'UV-A and UV-B' by producing phenolic compounds, UV-absorbing mycosporine-like amino acid (MAAs), or carotenoids (Chan et al. 2011).

Ethanolic, ethyl acetate and hexane extracts of *Sargassum polycystum* had skin whitening properties. Furthermore, antioxidants favor skin health by reducing hyperpigmentation (Bedoux et al. 2014). Topical fucoxanthin treatment prevented skin photoaging and wrinkle formation in UVB-irradiated human fibroblasts, possibly through antioxidant and antiangiogenic effects by reducing intracellular reactive oxygen species generated by UV-B radiation exposure (Heo and Jeon 2009). Astaxanthin and carotenoids are natural lipid-soluble pigments that directly protect the skin from UV light-induced photooxidation. Astaxanthin, derived from Haematococcus pluvialis, has been shown to improve skin wrinkles and texture while also providing effective protection against photooxidative damage. Studies have demonstrated that astaxanthin can protect against UV-induced damage in vitro. Additionally,

Table 3.2 Applications of macroalgae phycocolloids in the cosmetic industry (Pimentel et al. 2018)

Ingredient	Species	Technological function	Application	References
Carrageenan	Brown spp.	Thickening, gelling agent, blinder, sensory enhancer	Bath and shower gel	Pimentel et al. (2018)
Carrageenan	Brown spp.	Thickening, suspending agent, stabilizer, sensory enhancer	Oral, hair and skin care	
Alginate	Brown spp.	Interface vitalization	Shampoo	
Alginate	Brown spp.	Form retention	Dental moulds	
Alginate	Brown spp.	Emulsification, viscosity	lipstick	
Aminocyclohexene-type	Porphyra spp.	Anti-inflammatory	Skin care	Berthon et al. (2017)
Fucoidans, Fucans, galactan	Brown spp.	Anti-inflammatory, anti-aging	Skin care	
Phycobiliproteins	*Gracilaria gracilis*	Anti-melanogenic (whitening)	Skin care	

researchers have found that astaxanthin's photooxidation capabilities are stronger than those of lutein and b-carotene (O'Connor and O'Brien 1998). Phycobiliproteins have been used as pink and purple dyes in lipsticks, eyeliner, and other cosmetic formulations (Bedoux et al. 2014; Guillerme et al. 2017). Astaxanthin had a protective effect in the reduction of DNA damage and maintenance of cellular antioxidant status in lines of human skin fibroblasts, human melanocytes, and human intestinal Caco-2 cells, irradiated with UVA (Lyons and O'Brien 2002).

The phenolic content extracted from seaweeds is responsible for the mechanisms that protect cells from UV cytotoxicity and oxidative stress. Phlorotannins separated from four brown seaweeds have shown strong hyaluronidase inhibitory activity (Bedoux et al. 2014). Red seaweeds contain mycosporine-like amino acids, which are highly effective photoprotectors (Pandey et al. 2017).

3.2.2 Nutritional Application "Functional Food"

The role of macroalgae, particularly, in human nutrition and health, has been growing due to their low calories but it is rich in vitamins "A, B, C, E, and K" and essential minerals "Mg, Ca, Cu, K, Si, Zn, I, and Fe" which essential for human body's physiological functions in quantities more than vegetables and other land plants (El Zokm et al. 2021; Ismail 2022; Ismail et al. 2023b, 2025). They have the high ability to concentrate minerals from seawater and reach to 10–20 times more than mineral in land plants so they act as a significant source of valuable minerals for nutrition (Ismail et al. 2017; El Zokm et al. 2021; Ismail 2022).

3.2.2.1 Macroalgae as Human Food

The second Sustainable Development Goal is to end world hunger, provide food security, enhance nutrition, and advance sustainable agriculture. In the coming years, achieving SDG 2's aim of "Zero Hunger" will be a difficult endeavor due to the fact that the world population is expected to increase. Seaweeds are considered to be a healthy food for sustainable applications in the future due to their bioactive contents which can be used as a functional ingredient in many technological applications to obtain functional foods (Peñalver et al. 2020). They are also a source of a new class of high-quality protein, and they also contain some active ingredients not found in land plants (Chen et al. 2024). Seaweeds provide immediate health benefits and can be used as marine vegetables in human cuisine. They also have indirect value when used to extract phycocolloids, which are then added to meals. Some common edible macroalgae: *Sargassum*, *Porphyra yezoensis*, P. tenera "Nori", *Laminaria japonica* "Kombu", Undaria "Wakame," *Enteromorpha*, *Monostroma*, and *Caulerpa* for their flavour, texture, and culinary versatility. These species are frequently used in processed forms, such as flavourings for chips and snacks, as well as raw ingredients in salads, soups, and main dishes, like sushi. Approximately

80,000 tonnes of dried nori are produced during processing from Porphyra species, which are widely used sea vegetables. Production can reach 400,000 tonnes FW. Additionally, Undaria pinnatifida is eaten as a sea vegetable in a variety of forms that fall into two categories: traditional (suboshi wakame, haiboshi wakame) and new (judoshi wakame, kuki wakame)(Pérez-López et al. 2014). While, human consumption of brown algae (66.5%), red algae (33%) and green algae (5%) is high in Asia, mainly Japan, China and Korea (Polat and Ozogul 2013).

Edible seaweed species are an excellent source of proteins (> 48%) (El Zokm et al. 2021; Ismail et al. 2017, 2023a). About 50–60% polysaccharides which act as dietary fibers, especially soluble fiber such as alginates, which can influence satiety and glucose uptake from foods. These soluble polysaccharides may also act as prebiotics, stimulating growth of "beneficial" bacteria in the colon (Stiger-Pouvreau et al. 2016; Ismail et al. 2022). Moreover, they have very low in fats but contain essential fatty acids (FA) such as omega-3 and 6, which are regarded as essential FA since they must be obtained through food chains and cannot be biosynthesized by mammals. The ratio of ω6/ω3 of the studied seaweeds from Alexandria coaste was < 10 which endorsed them for nutritive application according to WHO which recommending this ratio should not be higher than 10 in diets (Ismail et al. 2016, 2025). Furthermore, they contain all the vital amino acids necessary for the body (Ismail et al. 2025).

The most common varieties of edible algae include: *Neopyropia/Porphyra/ Pyropia spp., Undaria pinnatifida, Saccharina latissima, Palmaria palmata and Chondrus crispus,* varieties that are associated with many health benefits, such as decreasing blood pressure, plus they are a valuable protein source (Probst et al. 2015). Interesting, nutrient contents of seaweed vary according to species, geographical distribution, season and environmental factors such as the water temperature, salinity, light and nutrients and mineral availability (Pena-Rodriguez et al. 2011; Ismail et al. 2025). *Corallina mediterranea* from Rhodophyta, *Padina boryana* and *Sargassum wightii* from Phaeophyta, and *Ulva lactuca* and *Enteromorpha linza* from Chlorophyta, were collected from Abo Qir, Alexandria. These seaweeds can be suggested as a source of nutritious food in Egypt based on appropriate ion quotient and estimated daily intake values (Ismail et al. 2017). Ismail et al. (2023b) recommended that the common brown species from Hurghada, Egypt promising candidates for nutritional applications. Macroalgal polysaccharides act as thickening, gelling agents, such as, ice cream and desserts. Like carrageenans, algainates and agars are employed as gelling agents and stabilisers for emulsions and suspensions, such as yogurt, health drinks, and even premium German beer. Their viscosity regulator acts as a fat substitute, ensuring silky textures and improving appearance, there by enhancing the quality of the product (Hernández-Carmona et al. 2013). Carrageenans play a crucial role in dairy products such as cheese, chocolate milk, and cocoa. They bind milk proteins, keeping milk solids suspended, preventing protein fractionation in cheese products, and allowing crystallization in milk ice creams (Ścieszka and Klewicka 2019). About 90% of the agar "vegetable gelatin," produced was for food applications and the remaining 10% were used for microbiological and other biotechnological uses (McHugh, 2003). Agar from *Gracilaria* and *Gelidium* species are the first hydrocolloids with European registration number E406 recognized as safe (GRAS) for use in food products. These hydrocolloids are considered safer alternatives. Flakes of *Palmaria palmata* and *Saccharina longicruris*, which are rich in protein, fiber, and minerals, were used to substitute salt and add edible fiber to Camembert-style cheese (Hell et al. 2018).

3.2.2.2 Macroalgae as Feed Source

The nutritional values of seaweed are rapidly gaining importance as a renewable source to substitute the conventional ingredients in the animal/aquaculture feed. Macroalgal biomass can be used as a feed supplement or as a replacement for conventional animal feed sources, based on comprehensive nutritional and toxicological studies. Approximately 30% of the world's current algal production is sold for use as animal feed (Khursheed et al. 2023). They can constitute a substitute for a protein source in the feed instead of soybean or canola meal or other nutrients, especially minerals (Shama et al. 2019). In the nineteenth and early twentieth centuries, there were many reports of occasional or systematic use of seaweeds to feed livestock in France (Brittany), the Scottish islands and Scandinavia (Gotland, Norway, Finland), mostly to ruminants (including calves) and pigs (Chapman and Chapman 1980).

Early at 1900s, seaweeds were preserved as silage and used as winter feedstuffs for sheep and cattle (Hallsson 1964. In times of scarcity, the Greeks collected seaweeds from the shore and washed them by fresh water, then gave them to their cattle and observed that these species prolonged their lives (Newton 1951).

They are included in the farm animal diet in the form of meal or as feed additive for animal nutrition "sheep, poultry, coat, dog, cat, and horse feeding" in the United States (Sathasivam et al. 2017). The addition of seaweeds to animal feeds results in the improvement performance, immunity of animal and meat quality (Sathasivam et al. 2017). Red seaweeds have garnered more attention because of their high ash content, especially in ruminant feed (Mahrose and Michalak 2022). Brown alga *Ascophyllum nodosum* is the most common species used in animal feed applications since because of its abundance, ease of harvest, and proximity to processing infrastructure (Evans and Critchley, 2014). *Palmaria* (*Rhodymenia*) *palmata* and *Alaria esculenta* were the most prefered seaweeds for sheep, while a frond of *Saccharina* (*Laminaria*) *latissima* was the suitable for horses (Hallsson 1964).

There is currently a growing interest in using unconventional feed resources, including seaweeds, to decrease the usage of antibiotics in animal agriculture. Seaweeds contain bioactive compounds and dietary fiber, which can improve intestinal integrity and decrease serum lipid concentration. Additionally, they promote immunological responses and reduce infections in farm animals (Cañedo-Castro et al. 2019; Mahrose and Michalak 2022). Algae added to the feed can solve this problem and additionally can improve the rumen microbiome and reduce the prevalence of foodborne pathogens (Zhou et al. 2018). Also, seaweeds can also be used to stabilize pH in the digestive system and provide favorable environment for microbial activity. For this reason, seaweeds rich in Ca and Mg are chosen—for example calcareous marine algae which contain high levels of calcium (300 g/kg) and magnesium (55 g/kg) (Cruywagen et al. 2015). Gastric juice pH in horses was stabilized with *Phymatolithon calcareum* and *Lithothamnion corallioides* (Jacobs et al. 2020). Especially seaweed polysaccharides are considered as an ideal prebiotic for animals. They stimulate the growth of beneficial bacteria, but inhibit the pathogenic strains (Kulshreshtha et al. 2014). Laminarin and fucoidan acting as natural antioxidants can enhance not only animal health, but also increase the quality of fresh meat (decrease in lipid oxidation) (Moroney et al. 2012). In addition to their prebiotic properties, seaweeds are able to maintain healthy gut flora and support gastrointestinal tract function, which in turn translates into strong immune system (Mahrose and Michalak 2022).

3.2.2.3 Macroalgae a Sustainable Aquaculture Feed Source

Since fish don't get all the nutrients they need from soybeans, seaweed can be used as a cost-effective and nutrient-dense alternative in fish meals (Chirapart and Ruangchuay 2022). They preferred as food by herbivorous fishes since their stomach had low pH levels and specialized guts required for the digestion of plant materials (Horn and Messer, 1992). The shape and texture of seaweeds determine their vulnerability to grazing by herbivorous fish. However, fleshy seaweeds are preferred over calcareous coralline and encrusting seaweeds (Ojeda and Muñoz 1999).

They have been declared as a "super feed", which has significant positive effects on fish performance due to their nutritive values, in addition to improving the immune system, antiviral, antimicrobial, improved gut function, stress resistance and and even ecological preservation (Cyrus et al. 2014). Also, they serve as an alternative for fish meal, because plant-based proteins do not contain such high P levels, and it would help to take the pressure off wild fish stocks (Ayoola and Idowu, 2008).

The use of seaweed species as a component of aquafeed for various fish species has been the subject of increased investigation. Regarding this, Lomartire et al. (2021) discovered that incorporating seaweeds into fish diets enhanced many aspects of fish species' physiological activity, growth, lipid metabolism, carcass quality, disease resistance, and stress response. Abdel-Aziz and Ragab (2017) discovered that using the green macroalgae *'Ulva* and *Enteromorpha'* as half of the feeding rate with artificial feed improved rabbit fish fry growth and reduced feed costs, whereas replacing artificial feed with fresh seaweeds had a negative effect on rabbit fish fry growth. The inclusion of brown algae in fish meal improve growth, feed intake and feed texture, reduce dry matter loss and increase water absorption, proving to be an excellent binder for pelletized feeds (Cruz-Suárez et al. 2009).

Macrocystis pyrifera and *Undaria pinnatifida* water extracts as feed additive to *Artemesia longinaris* diets improved the growth of *Artemesia longinaris* (Díaz et al. 2017). The fresh *P. palmata* is the most efficient food for the growth of mollusk in farming conditions (+107.8% weight gain) (García-Bueno et al. 2014). In order to reduce the production costs of aquatic feed, Manikandan et al. (2022) suggested replacing 50% of fish meal with marine red seaweed (*Halymenia dilatata*). This substitution was found to significantly enhance the growth performance, chemical body composition, and digestive enzymes of fish *Labeo rohita.*

Interestingly, seaweeds act as the major market for astaxanthin and as pigmentation source in aquaculture (Canales-Gómez et al. 2010). In 1987, the United States Food and Drug Administration approved the use of astaxanthin as a feed additive for aquaculture and subsequently in 1999 astaxanthin were be approved as a nutraceutical. It was the most important carotenoid in salmons and rainbow trouts (Tolasa et al. 2005).

3.2.2.4 Macroalgae as Supplement for Broiler Feed

Broiler feed is based primarily on corn and soybean meal (SBM) are the based primarily feed of chicken. Algae are a natural source that can improve the eggs quality by increasing the level of total carotene, lutein, and zeaxanthin, as well as the antibacterial and antioxidant contents (Wang et al. 2013a, b; Evans and Critchley, 2014). In general, algae can be safety used as replacement for conventional proteins up to a level of 5–10% without affecting egg production quality attributes or broiler performance (Al-Harthi and El-Deek 2012). However, long-term feeding of algae at higher concentrations may have negative consequences. Utilizing several seaweed species can improve the weight, quality, and yolk cholesterol of chicken eggs; other seaweed-derived macromolecules can lower the amounts of dangerous bacteria in poultry (Farghali et al. 2023).

Macroalgae has been used in poultry to improve their immune status, decrease microbial load in digestive tract, and for their beneficial effect on quality of poultry meat and eggs (Abudabos et al. 2013; Wang et al. 2013a, b). When comparing birds that were fed 3.0% U. lactuca to those that were fed the control or 1.0% *U. lactuca*, there was increase in dressing percentage and breast muscle yield (Abudabos et al. 2013).

Wang et al. (2013b) recorded that addition of *E. prolifera* in broiler feed at 1–3% resulted in improved egg production and quality, yolk colour, reduced cholesterol in yolk, increased the body weight, shell thickness and it also resulted in a lower *Escherichia coli* load in faeces, suggesting better animal health. *Ascophyllum nodosm* had no efficiency on body weight, but it positively influenced egg yolk color and growth performance (Evans and Critchley 2014). *Sargassum* species which collected from the Red Sea shore exhibited no deleterious effect on body weight, egg weight, egg production, feed conversion ratio and egg quality when inclusion in hens diets at 1–12% during 20–30 days (Al-Harthi and El-Deek 2012). Especially, calcified red seaweeds are a valuable alternative source of calcium for broilers, since the availability of organic calcium is higher than that of inorganic calcium from mineral sources such as limestone (Bradbury et al. 2012). Seaweed-derived sulfated polysaccharides have the potential to enhance immune response, antioxidant status, and small intestine ecology. They can also promote the growth of good gut microbes like lactic acid bacteria while inhibiting the colonisation of dangerous bacteria like *Salmonella enteritidis* and *Escherichia coli* (Li et al. 2018). Recentaly, a novel approach in animal feeding was proposed by El-Maaty et al. (2021) Brown alga *Sargassum latifolium* converted into extract was used for the production of calcium nanoparticles supplemented to the diet of laying hens, which can improve egg quality, including eggshell thickness and shell weight. Red seaweeds *Chondrus crispus* and *Sarcodiotheca gaudichaudii* supplemented to the diet of laying hens were effective in providing protection against *Salmonella* Enteritidis colonization, which ensured safer egg production and minimization the spread of salmonellosis to humans (Kulshreshtha et al. 2017). Green and brown seaweeds supplementation can improve egg-laying performance and had a positive effect on fertility and hatchability values and resulted in a significant reduction in total lipids and total cholesterol in the serum and yolk while enhancing total antioxidant capacity in blood serum as well as positively affecting economic parameters (Abu Hafsa et al. 2019).

3.2.3 *Industrial Applications*

3.2.3.1 Hydrocolloids Substances

In addition to being value as food and traditional remedies in their natural state, seaweeds have been actively utilized and promoted by various businesses due to their exceptional hydrocolloid qualities which are frequently utilized in food applications such as bakery, dairy, meat, and milk products (Abdul Khalil et al. 2018). The addition of seaweed components or thallus as important compounds improves the nutritional value and texture of food products while offering protection against a number of chronic illnesses, including diabetes, hypertension, obesity, and dyslipidaemia (Sharma and Baskaran 2021).There have been report of red and brown seaweeds being utilized as food ingredients at quantities of 0.5–8% in bread, 2.5–20% in cake, 5–60% in biscuits, 3–9% in cookies, 3–30% in noodles, 5–20% in pasta, and 3.5% in extruded maize (Sharma and Baskaran 2021; Quitral et al. 2022). Alginate, agar, and carrageenan are three outstanding hydrocolloid components from seaweeds frequently used as stabilizers, thickeners, emulsifiers, and fillers in the food, pharmaceutical, cosmetic, textile, and biotechnology industries (Makkar et al. 2016). Alginates from brown macrolagea are commercially important due to their gelling and biological abilities. They are natural, biocompatible, biodegradable, bioadhesive and nonimmunogenic. It is commonly used in the pharmaceutical industry in gastric alkalis, binding agents for tablets, wound dressings, and dental impressions (Leslie et al. 2013). Additionally, it has many industrial applications such as textiles, cosmetic, electrodes and the food industry as stabilizers and thickeners in e.g., products like jelly, drinks, and desserts (Rhein-Knudsen et al. 2015).

Commercial carrageenans are extracted from the red macroalgae "*Chondrus, Eucheuma, Gigartina, Hypnea,* and *Kappaphycus,*", which comprise up to 50% of the dry weight (Van De Velde et al. 2001). Three kinds of carrageenan were isolated from different species of algae, each possessing unique characteristics, e.g. κ; ι and λ-Carrageenan. κ-Carrageenan is predominantly produced from *Kappaphycus alvarezii*, while ι-carrageenan is mostly extracted from *Eucheuma denticulatum*, and λ-Carrageenan is extracted from genera of *Gigartina* and *Chondrus* (Van De Velde et al. 2001).

Carrageenans are used in various drug delivery systems as matrixes to control drug release, microcapsules, microspheres, binders in toothpaste, thickeners and stabilizers in cosmetics. They are also used as stabilizers, gelling agents, emulsifiers, and thickeners in the food and baking industries (such as ice cream, cheese, jam, and bread dough). The global production of carrageenans was approximately 60.000 tons/year, with an estimated value of around US$ ~ 626 million in 2014 (Rhein-Knudsen et al. 2015).

Agar is extracted from the agarophytes red seaweeds, e.g., *Gelidium*, *Gracilaria*, and *Gelidiella*, and is the raw material for agar production (Mchugh 2003). It can act as a good bio-medical impression material in dentistry, due to its thermo-reversible properties; it melts and liquefies at high temperature and sets to gel while cooling. Additionally, it is utilized to create pharmaceutical-grade growth media for laboratory purposes. In the USA, approximately one-sixth of agar is employed as a culture medium because of its low metabolite rate, non-digestibility, gel firmness, elasticity, clarity, and stability (Abdul Khalil et al. 2018).

In particular, adding the polyphenols from seaweed to meat products improves the meat's antioxidant capacity and prolongs its shelf life. This is accomplished without requiring people to alter their eating habits by postponing the oxidation of lipids in muscle tissues and boosting the amount of protein in their diets (Wang et al. 2023).

3.2.3.2 Natural Dyes

Synthetic dyes are toxic and harmful to all living things. Consequently, there is a growing global demand for natural dyes due to their safety, environmental friendliness and unique shades. Their use should be considered a superior alternative to synthetic colors. Seaweeds contain numerous original pigments that are not found in terrestrial plants, characterized by solubility and safety. In this context, *Caulerpa lentillifera* can be utilized as a natural dye source, producing unique and interesting shades on textiles (Abd Kadir et al. 2014). The ethanolic or aqueous extract of brown alga *Sargassum* is a valuable source of dye for coloring leather (Venkateswarlu 2015). Also, the dye extracted from *Sargassum myriocystum* has potential application as a food colorant (Annam Renita and Davuluri 2015). Additionally, dyes extracted from *S. thunbergii*, *S. horneri* and *Hizikia fusiforme* have significantly higher sun protection factor, antibacterial properties, and deodorization compared to dyed cotton, wool, and silk (Bak 2010).

R-Phycoerythrin "R-PE" is derived from *Porphyra* sp. (nori) and is utilized as a food colorant in many Asian countries. Particularly, the α R-phycoerythrin subunit of *P. yezoensis* has the potential to be used as a natural colorant in the food and cosmetics applications (Ulagesan et al. 2021). Phycobilproteins have become more crucial to replace the synthetic colours due to their medical effects (Ismail et al, 2023b).

3.2.3.3 Antifouling Substances

Biofouling is a natural phenomenon of marine ecosystem occurred by the surface colonization and development of micro and macrofoulers on submerged natural/artificial substratum, leading to huge economic and environmental losses worldwide such as roughness in the ship's hull, reducing efficiency, speed and maneuverability, aquaculture nets and pilings, and causing an increase 35 to 50% in fuel consumption of a ship that has spent 6 months at sea. Most antifouling techniques have depended on organotin (tributyltin) or heavy metals (copper, zinc) based paints that have abroad spectrum toxins to target and non-target marine organisms which lead to serious environmental problems at low concentrations as low (ppb) such as killing organisms and accumulation in the environment. According to the International Maritime Organization, many harmful antifouling compounds such as TBT (tributyltin) have been banned for marine applications (Satheesh et al. 2016).

There were more than 160 research publications about antifouling chemicals from marine algae between 2010 and 2017, most of which concentrated on green algae and fewer on brown and red algae (Dahms and Dobretsov 2010; Ismail and Amer 2020). Ismail and Amer (2020) recommended that sulfated polysaccharides can be used as a natural antifouling agent against the bacterial biofilm which is the base layer of the fouling process. The antifouling ability of seaweeds related to their phytochemical contents as explained in Table 3.3 (Dahms and Dobretsov 2017). Interesting, most of these substances are easily biodegradable and nontoxic than synthetic antifoulants (Boyd et al. 1998). The ethanol extracts of *U. reticulate* exhibited antibacterial activity that correlated with the major functional groups present, including amino, carbonyl, hydroxyl, phosphoryl functionalities, aliphatic (fatty acids), and NH_2 groups. This activity was observed against *Flavobacterium* sp., with minimal efficiency toward the remaining three biofilm bacteria (Dahms and Dobretsov 2017). In last years, a total of 100 compounds from both *Laurencia obtusa* and *Acanthophora spicifera* were computationally screened against *Mytilus galvoprovinces* (4CN8) and *Megabalanus rosa* (6LEK) target proteins. Thyrsiferol compound from *L. obtusa* exhibited the maximum dock score (-8.6 kcal mol^{-1}), and Cholestaa-6, 22, 24-Triene,4,4-Dimethyl compound from *Acanthophora spicifera* showed the highest dock score (-8.9 kcal mol^{-1}) against macro fouling protein with low toxicity effects (Suresh et al. 2024). The green alga *U. lactuca* produced hexadecanoic acid as one of the promising compounds from the seaweed group as an eco-friendly antifouling agent (Sukrri et al. 2024).

Antifouling compounds from macroalgae have been isolated and tested in an ecologically-relevant way (Dahms and Dobretsov, 2017). In this connection, Salama et al. (2017) demonstrated that the methanolic extracts of *Ch. linum*, *T. ornata* and *S. polycystum* from the Red Sea in Saudi Arabia significantly inhibited the settlement of cypris larvae of the common fouling barnacle, *Amphibalanus amphitrite* in vitro. They also reduced biofouling growth on nylon net panels in the field to varying degrees over three months as explained in Fig. 3.5.

da Gama et al. (2008) stated that red macroalgae are the major producers of antifouling defenses. In this context, Salem et al. (2020) observed that *J. rubens*

Table 3.3 Antifouling compounds from different Macroalgal groups (Dahms and Dobretsov 2017)

Macroalgae species		Compounds	Biological activities
Chlorophyta	*Chlorococcum humicola*	Chlorophyll a and b	Antibacterial andAntifungal
	Caulerpa prolifera	Acetylene sesquiterpenoid esters	Antibacterial and Antialgal
	Ulva sp.	B- carotene	Antifouling
	U. pertusa	Alkaloid phenolic acid	Antialgal
Phaeophyta	*Dyctiota* spp.	Diterpenes	Antibacterial
	Fucus vesiculosus	Dimethylsulphopropionate Proline	Antibacterial
	Sargassum sp.	Phlorotannins	Antialgal
	S. muticum	Galactoglycerollipids	Antibacterial
	Taonia atomaria	Sesquiterpenoids	Antibacterial
Rhodophyta	*Asparagopsis taxiformis*	2-dodecanoyloxyethanesulfonate	Antibacterial and Anti-QS
	Galdieria sulphuraria	Floridoside	Antifouling
	Laurencia translucida	Fatty acid derivates	Antibacterial
	L. translucida	Fatty acid	Antifouling
	L. viridis	Chlorophyll a and b	Anti-diatom

showed the highest antibiofilm activity compering with *S. vulgare*, *U. fasciata*. *Delisea pulchra* was found to be active against the settlement of macrofouling larvae, bacterial colonization, and spores of fouling algae (Kjelleberg et al. 1997). In contrast, Blunt et al. (2011) reported that brown algae have a higher antifouling activity compared to red and green seaweeds, specifically, the families Sargassaceae and Dictyotaceae.

The antifouling activity of seaweeds is linked to epiphytic bacteria that repel other fouling bacteria, thus protecting the seaweed from fouling by other organisms (Bhosale et al. 2002). The majority of bacteria isolated from *Colpmenia sinuosa* produce Quorum sensing (QS) (Kanagasabhapathy et al. 2009). Harder et al. (2004) isolated *Vibrio* sp. from *U. reticulate* that produced antifouling compounds.

Interestingly, some calcareous red seaweeds (family Corallinaceae) have a unique physical antifouling mechanism of epithallus sloughing toward epibionts (Villas-Bôas and Figueiredo, 2004). The arborescent red alga *J. rubens* is described as an alga strongly impregnated with calcium carbonate and its extract has shown strong antifouling activity towards mussels, supporting the idea of a chemical defense mechanism in this alga which is already physically defended (Joly, 1967). The non-toxic green synthesized Fe_3O_4-NPs may be suitable for development of new antifouling marine paints. They can also be used in marine paint formulations instead of iron oxide because they are more efficient and safer (Salem et al. 2020).

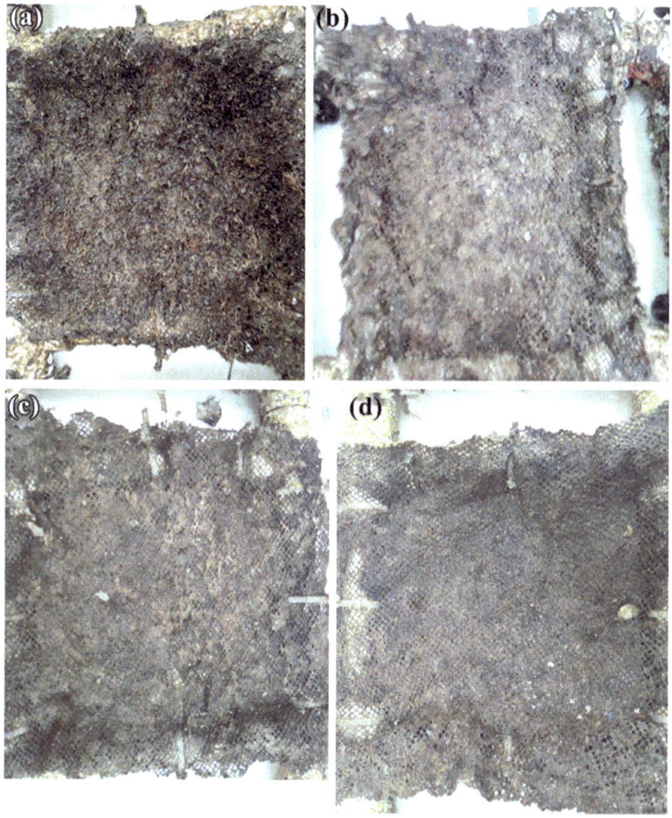

Fig. 3.5 Biofouling growth on nylon net test panels coated with macroalgal extracts submerged in the sea for three months. **a** Control, without any coating. **b** *Ch. linum* extract. **c** *T.* ornata extract. **d** *S. polycystum* extract (Salama et al. 2017)

3.2.3.4 Paper Production

Countries with limited wood resources have been searching for alternative sources that can produce higher quality end products and reduce paper manufacturing costs. Algae, known have emerged as a potential source of cellulose along with terrestrial plants. In particular, marine seaweeds have gained popularity as a source for producing Cellulose nanocrystals (CNCs) due to several advantages (i) easy extraction process due to low levels of other interfering polysaccharides; (ii) seaweed have a higher yield of stored carbohydrates; (iii) their rapid growth allows for quicker harvest compared to other sources (Salem and Ismail 2021). Seaweeds used for paper manufacturing, mainly green algae, are known to contain more cellulose than red and brown algae, at a similar level to wood (Salem and Ismail 2021). The significant amount of cellulose found in green algae *Ulva* sp. a long with its low percentage of lignin-like compounds makes it an interesting raw material for partially replacing wood pulp in the production of pulp and paper (Salem and Ismail, 2021). Seaweed waste contains cellulose, the main raw material for making pulp which is the raw material for paper, which giving them great importance for their use in obtaining paper (Hasna and Junianto 2021). Additionally, alginates are commonly used in paper production (Radmer 1996). Seo et al. (2010) found that handsheets produced from red algae *Gelidium amansii* and *Gelidium corneum* pulp had high Bekk smoothness, brightness, and opacity, essential properties for high-value printing paper, compared to those of wood pulp (Table 3.4).

In addition, the fiber content of the red algae handsheet is short and very narrow as shown in Fig. 3.6. Recently, blending seaweed and coconut fiber has allowed for customising biopaper properties, such as density and flexibility, making it suitable for a broader range of applications and encouraging environmental sustainability (Khan et al. 2025). Numerous benefits have been reported from the use in papermaking. It helps avoid dependence on tree monocultures and the variability of the price of wood sources, which are expected to increase over time. Furthermore, it involves low consumption of chemical reagents due to its easy treatment, valorization of coastal waste associated with environmental and economic problems in tourist areas, and higher productivity of biomass than non-fertilized crops (Baweja et al. 2016).

3.2.3.5 Biofuel Applications

One of the seven SDGs is the production of bioenergy, and sustainable biomass feedstock sources are necessary to fulfil the growing demand for oil and the depletion of petroleum reserves. The third generation source of biofuel (bioethanol, biodiesel, and biogas) is algae biomass (Fig. 3.7) and could provide a lion's share as an alternative to the common plant crops derived biofuels (Ismail et al. 2020a, b). Third-generation biofuels focus on algae and cyanobacteria, whereas fourth- generation biofuels involve genetic engineering to increase desired traits of organisms used in biofuel production. Seaweed biofuel has the potential to lessen greenhouse gas emissions, lessen dependency on finite fossil fuels, and promote economic growth in coastal areas by using the enormous resources offered by marine ecosystems. Pluse to their environmental benefits, including the capacity to absorb carbon, reduce land usage, and be carbon–neutral (Olanrewaju et al. 2024, Ismail and El Zokm 2025).

The majority of research has focused on creating fuels from microalgae rather than macroalgae, as microalgae have a higher lipid content. Seaweeds represent a promising and sustainable feedstock for biofuel production which raises increasing research interests. Howard Wilcox presented seaweed as a potential energy source in the late 1960s, envisioing their production as a solution not only to the energy crisis but also to global warming (Wilcox 1975). This is in contrast to the extraction of oils, which make up less than 5% of the total dry weight that and can be processed into biodiesel or hydrocarbons. Additionally, they reduce greenhouse gas emissions through the photosynthesis process (Ismail et al. 2020a, b; Ismail and El Zokm 2025).

Table 3.4 Comparison of handsheet properties (bleached chemical wood pulp and red algae pulp (Seo et al. 2010))

	Term	Caliper (l m)	Basis weight (g/m²)	Density (g/cm³)	Breaking length (km)	Stretch (%)	Smoothness[a] W/F (s)	Drainage (s)	Brightness (%)	Opacity (%)
Wood pulp	WP	92.3	60.02	0.65	4.70	2.94	4.2/6.3	6.96	83.68	77.10
G. corneum	GC-120	106.5	61.08	0.57	4.04	2.60	43.5/65.4	14.85	86.59	92.81
	GC-140	113.3	63.28	0.56	3.74	3.02	38.5/82.4	18.41	81.01	96.79
	GCS-120	80.0	53.45	0.67	4.17	2.46	42.3/68.6	18.88	86.03	92.30
	GCS-140	108.3	60.60	0.56	3.41	1.83	38.8/83.2	18.13	83.71	96.08
G. amansii	GA-120	124.3	62.17	0.50	2.89	2.79	52.3/77.8	9.79	82.47	92.85
	GA-140	130.7	60.30	0.46	2.98	3.46	44.3/68.4	10.01	77.97	95.72
	GAS-120	126.8	60.98	0.48	3.25	4.52	53.2/85.3	10.74	86.66	95.51
	GAS-140	141.5	63.73	0.45	2.93	3.47	45.7/74.2	10.73	84.00	95.39

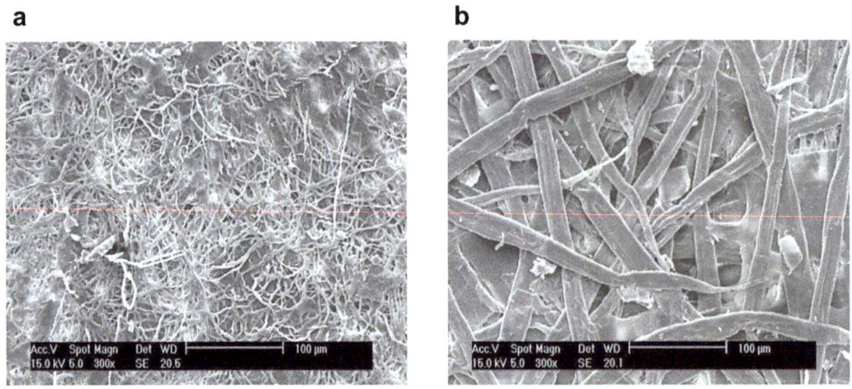

Fig. 3.6 Shape of *Gelidium* pulp (**a**) and Wood pulp in handsheet (**b**) (300 X) (Seo et al. 2010)

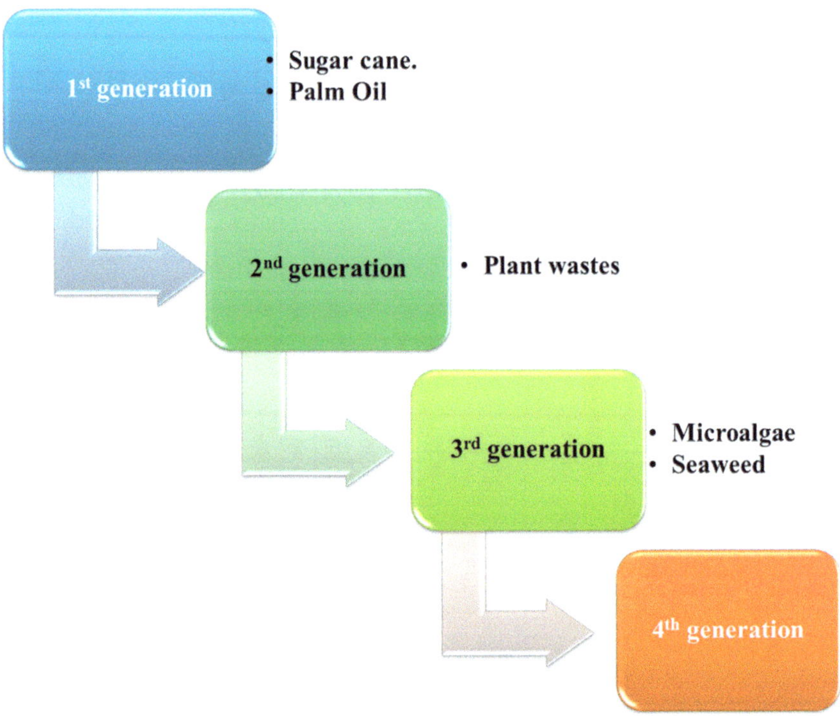

Fig. 3.7 Evaluation of biofuel (Own elaboration)

Bioethanol Production

During the oil crisis in the late 1970s several research groups, predominantly in the United States under the "US Marine Biomass Energy Program" (1968–1990), were looking at seaweed for mass cultivation and conversion to bioethanol (Chynoweth 2001). Seaweeds have several advantages over lignocellulosic plant biomass. Their polysaccharides are easily hydrolyzed and fermentable because to their high carbohydrate content (20–75% of total cell mass), higher biomass production, and lack of lignin (Yanagisawa et al. 2013). Additionally, the leftover algal cake is high in proteins after the starch and lipid fractions are extracted for the generation of bioethanol and biodiesel. These proteins can be utilized to create useful byproducts, such as fish and poultry feed additives (Ismail et al. 2020a, b; Ismail and El Zokm 2025).

There are many macroalgal species had the ability to accumulate high levels of carbohydrates, which are suitable as a substrate for microbial conversion processes and production of bioethanol (Holdt and Kraan, 2011). Regardless of the species of seaweed utilized, the ultimate ethanol yields could only reach about 15–20 kg of pure ethanol per tonne of seaweed biomass. The concentrations of released glucose only matched about 3–4% of the initial seaweed biomass' dry weight. Furthermore, seaweed contains other sugars besides glucose, such as galactose, xylose, and mannitol, which, when properly utilized and fermented by the right microorganisms (like *Zymobacter palmae*, *Pichia angophorae*, and engineered strains like *Escherichia coli* EO11 and *E. coli* BAL1611), could yield more bioethanol per tonne of seaweed (Owusu et al. 2024). The high potential growth rates and high carbohydrate content of *Ulva* and *Enteromorpha*, which can make up 38–54% of cell wall polysaccharides on a dry weight basis, have made them viable energy crops (Lahaye et al. 1997). Ismail et al. (2020a, b) recommended *U. linza* as a costless candidate for bioethanol production. Owusu et al. (2024) stated that within the framework of a circular economy, invasive *Sargassum* species present themselves as promising candidates for bioethanol production. Huesemann et al. (2010) found that laminarin and mannitol derived from *Laminaria hyperborean* gave high yields to ethanol by yeast, *Pichia angophorae.*

The physical, chemical or biological pretreatments (hydrolysis) are the solvent to enhance the ethanol production from seaweed, which disrupts the algal cell walls and leads to increasing the saccharification efficiency (Truus et al. 2001).

Saccharification and fermentation can be also carried out simultaneously using an amylase enzyme-producing strain for the production of ethanol in a single step (Schenk et al. 2008). Saccharomyces cerevisiae is the primary decomposer used in industrial ethanol production through fermentation. It has a strong ability to survive and function at ethanol concentrations of up to 10% before breaking down, potentially causing the fermentation process to halt (Yusaf et al. 2011).

Utilizing complex carbohydrates as a feedstock for the production of bioethanol is becoming simpler thanks to new biotechnologies based on microbial enzymes (Milledge et al. 2019). *Gelidium elegans* was effectively used by Yanagisawa et al. (2013) to produce a higher concentration of ethanol (55.0 g/L) than *U. pertusa* and *L. japonica.* Ethanol is recovered from the fermentation broth by distillation, which is the most energy-intensive step in the ethanol production process. A 4–5% ethanol production concentration is required to make the process commercially viable; however, this ratio has not yet been reached by converting solely seaweed glucans.

Biodiesel Production

Bio-oils present an alternative to liquid biofuels with similarities to petroleum oil (Kishimoto et al. 1994). Microalgae are considered as a sustainable source of biodiesel which has the ability to synthesize and accumulate significant quantities of lipids as they are sunlight-driven oil factories via transesterification process the algal oil converts into biodiesel (Abomohra et al. 2017). Seaweeds seem less suitable for bio-oil production due to their low lipid content, high water, nitrogen and ash content, as well as the presence of metals (Suganya et al. 2013).

The yield of bio-oil from some macroalgae hydrothermal liquefaction reached to > 23% with an energy recovery from the original macroalgal biomass of 63% for *Laminaria saccharina*, and 56% for *E. prolifera* (Murphy et al. 2013). *Padina minor* showed the highest potential in terms of its lipid content and it may be a suitable candidate for biodiesel production (Rani and Kumar 2021). For possible commercial applications, the green alga *Caulerpa prolifera* is a dependable biodiesel generator with an attractive oil composition ratio (Ismail et al. 2020a, b). The extracted bio-oil from *Sargassum* would be healthier and more environmentally friendly than fossil fuels due to zero SOx and NOx emissions (Rahbari et al. 2019).

Biogas Production

Biogas is a mixture of gases composed mainly of methane, carbon dioxide and small quantities of hydrogen sulfide from organic matter which is commonly produced by anaerobic digestion processes (Irfan et al. 2019). As compared to terrestrial ligno-cellulosic biomass, seaweeds are rich in polysaccharides, contain no lignin, and have little cellulose content, making the biomass amenable for biogas production through anaerobic digestion (Laurens and Chen-Glasser, 2017). The potential of algal biomass as a source of liquid and gaseous biofuels has been the subject of considerable research over the past few decades, with researchers strongly agreeing that algae have the potential of becoming a viable aquatic energy crop, with a higher energy potential compared with that of either terrestrial biomass or municipal solid waste (Milledge et al 2019). Seaweeds are considered a suitable substrate for biogas production owing to their high carbon–nitrogen ratio, low lipid content and lack of lignin. The huge amounts of stranded biomass represent an attractive feedstock for energy production and could be integrated into a biorefinery scheme (Zhao et al. 2022). Huesemann et al. (2010) reported that Tokyo Gas studied the production of biogas from macroalgae that was collected from biomass naturally deposited on beaches after storms and high tides.

Biomethane Production

Macroalgae acts as a source for biomethane production but the production cost was 2–10 times higher than the cost of natural gas (Carlsson et al. 2007). Pokoo-Aikins et al. (2010) concluded that the overall costs of producing substitute natural gas from marine macroalgae are higher than those of grass, wood, waste systems and substantially higher than fossil fuels in the U.S. Biomethane yields depend mainly on the fermentation conditions and pretreatment of algae " with heat and alkali' which may increase the gas yields (Mendis and Kim, 2011).

Methane yields from green seaweeds were higher than yields from brown and red seaweeds (Hansson, 1983). Chynoweth (2001) suggested *Ulva* species to produce methane gas since its easily digested to methane gas and seem to lack epiphytes, which can interfere with their production and also do not contain lignins. Gurung et al. (2012) found slightly higher methane yield from brown algae '6.14 MJ of energy per kg of dry weight' during anaerobic digestion. *Gracilaria* sp. and *Cladophoropsis ystis* (kelp) were excellent marine algae for biomethane generation (Bird et al. 1990). Hanssen et al. (1989) screened the biogas production from *L. hyperborea*, *L. saccharina* and *A. nodosum* and found maximum methane yield of 0.28 CH_4/g VS during semicontinous cultures of *L. hyperborea.* Hansson (1983) increased the methanol yields (350–480 ml CH_4 g/ VSdd) from *Ulva*, *Chaetomorpha* and *Cladophora* by the fermentation process. High values of biomethane potential (BMP) have been obtained from the brown seaweed *Macrocystis* (0.39–0.41 m3 CH_4/kg Volatile solids (VS)) (Singh and Gu, 2010), the red seaweed *Gracilaria* (0.28–0.4 m^3 CH_4/kg VS), these values are comparable to the BMP of terrestrial crops such as sorghum (0.26–0.39 m^3 CH_4/Kg VS), sugarcane (0.23–0.3 m3 CH_4/Kg VS) (Chynoweth, 2005). The brown alga "*Sargassum casifolium*" has the potential to be used as biogas material "methane 24.56%" (Irfan et al. 2019). The hydrothermal and thermo-acidic pretreatment were improved the CH_4 yield of some macroalgae species (Thompson et al. 2019). Thermo-acidic, hydrothermal and alkaline pretreatments were evaluated for improving the CH_4 production of beach-cast seaweed (mixture of *Pilaiella littoralis*, *Ectocarpus siliculosus* and *Zostera marina*) through anaerobic digestion. The thermo-acidic pretreatment resulted into the highest release of sugars and the highest increase of CH_4 yield (78% compared to raw seaweed). A preliminary techno-economic assessment indicated that the alkaline pretreatment permits more space for improvement on the cost of CH_4 production of seaweed compared to the thermo-acidic pretreatment (Lymperatou et al. 2022). Pretreated macro-algae biomass using 0.2 M HCl at 80 °C for 2–24 h, resulting in 39–140% increases in biomethane yields (Barbot et al. 2015).

Biobutanol

Butanol (n-butanol) is known as a superior fuel substitute for gasoline, outperforming ethanol. It has also been used as a profitable chemical feedstock in various industry sectors. The two steps of biobutanol production are acidogenesis and solventogenesis (Mahapatra and Kumar 2017). In the United States, research and investment are currently focused on biobutanol obtained from macroalgae. This is due to its ability to be produced on a pilot scale using sugars from algae (Huesemann et al. 2010). *Ulva lactuca* was a substrate for acetone, butanol, and ethanol fermentation (Van der Wal et al. 2013). The yield of butanol from fermented *U. lactuca* by *Clostridium* strains (+ve) is lower than that of ethanol produced under comparable conditions (Nikolaison et al. 2012). Huesemann et al. (2010) found that major improvements are still required to make industrial-scale butanol production from macroalgae economically viable.

Biohydrogen

Hydrogen can be considered a future fuel as it may become a significant factor in maintaining energy balance (Hassan et al. 2024). It is a clean and renewable energy carrier with a high energy density (122 MJ/Kg). Regarded as one of the promising fuels of the future, it represents a scientific, environmental, and ecological challenge (Pierra et al. 2014). Furthermore, the coupling of biohydrogen production with seaweed farming can be a win–win solution for both energy and environment. After 15 days of anaerobic fermentation, the sample treated with sonic surfactant disintegration (SSD) (154 mL H_2/g COD) produced the highest amount of biohydrogen from *Chaetomorpha antennina* compared to the control (45 mL H_2/g COD) and sonic disintegration (SD) samples (Shabarish et al. 2023). Some countries like Japan and South Korea have already been leading initiatives to produce hydrogen using seaweed biomass within the wider context of energy transition actions (Harichandan 2025). A combined steam heating and dilute acid pretreatments to improve the solubilization of *Gelidium amansii*, increasing the biohydrogen yield by 92.6% (Sivagurunathan et al. 2017). Liu and Wang (2014) pretreated *Laminaria japonica* using steam heating at 121 °C for 30 min, increasing the fermentative biohydrogen yield from 10.0 to 66.7 mL/g total solids.

3.2.4 Agriculture Applications of Macroalgae

Seaweed and their extracts are growing in popularity in agriculture due to their advantages over chemical fertilizers. These benefits include a longer shelf life, ease of handling, increased soil fertility through humus formation, and lower cost. They considered as proming multifunction bioinoculants and ecofriendly environmental tools in organic farming. They serve as an exceptional natural biosource in agriculture by enhancing the physicochemical parameters of soil, producing inducing substances (natural biofertilizers), and biocidal substances against plant infecting pathogens due to their bioactive compounds which act as biofertilizer, biostimulator, soil conditioner, nematicide, insecticide and microbicidal effects as depicted in Fig. 3.8 (Battacharyya et al. 2015; Prisa et al. 2024).

Furthermore, seaweeds can promote plant growth in cold, drought, and salt conditions, increase disease resistance, and enhance agricultural production and productivity (Hofmann et al. 2018). Extracts have recently been shown to be effective

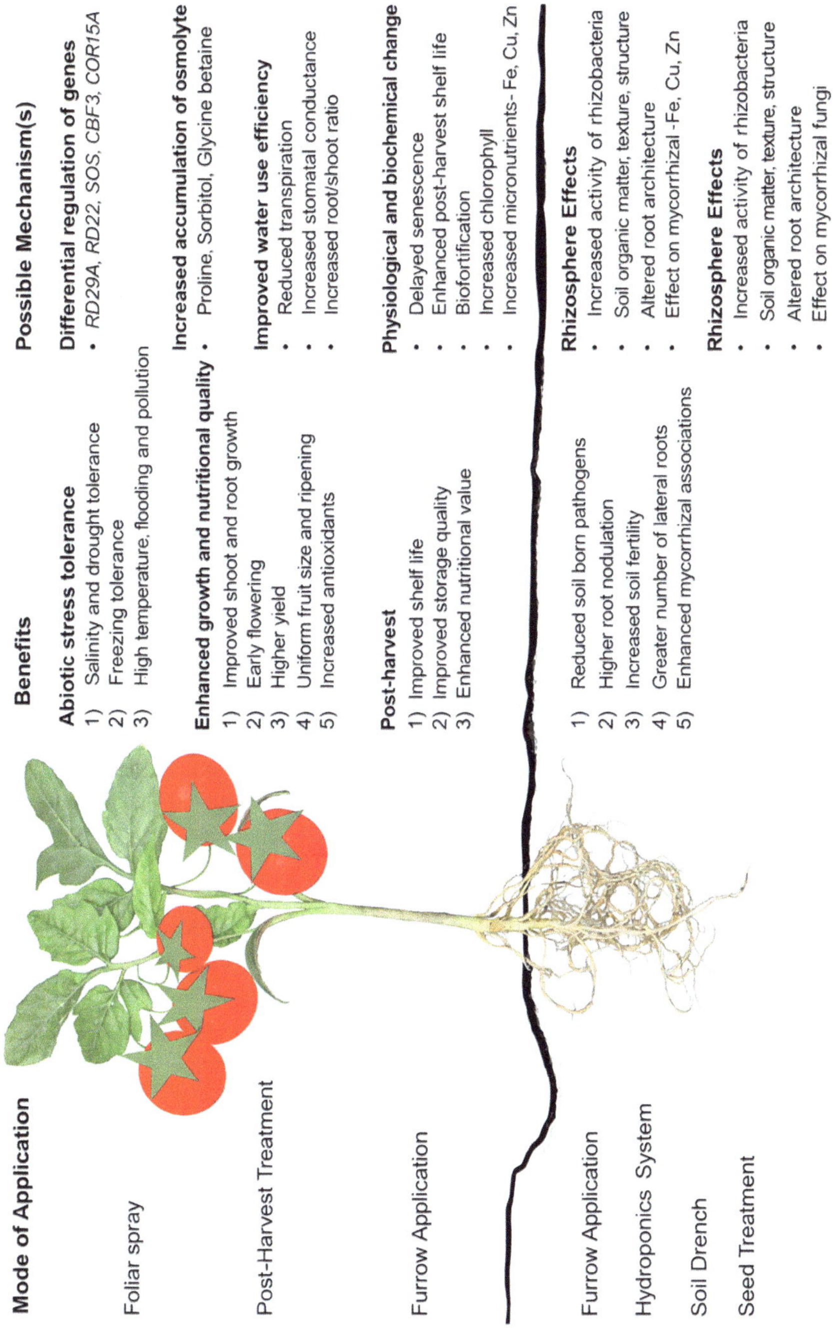

Fig. 3.8 Schematic representation of beneficial effects of algal extracts and possible mechanisms of bioactivity (Battacharyya et al. 2015; Prisa et al. 2024)

in promoting growth in various specialized uses, such as hydroponic methods and in vitro tissue culture (Prisa et al. 2024). In addition to lowering the use of harmful pesticides, the use of algae extracts in agriculture helps to protect the environment. Algae extracts have the potential to significantly increase food yields when applied into agricultural practices globally, particularly in light of the anticipated rise in the world's population by 2050.

3.2.4.1 Macroalgae as Biofertilizers and Soil Improvers

Seaweeds have been used to fertilize fields since ancient times. They have been a valuable resource for many years, being exploited to fertilize agricultural soils in coastal areas. Dried or fresh seaweed help to enhance biochemical constituents of the plant e.g., carbohydrates, lipids, proteins, fibers, ash, phenol, dietary fiber fruit yields, photosynthetic activity, nucleic acid synthesis and ion uptake and increase the yield and productivity of several crops, germination rate of seeds (Ismail and El-Shafay, 2015; El-Sheekh et al. 2016). This could be attributed to the high content of macroelements (Ca, Mg, K, and P), microelements (B, Co, Cu, Fe, Mn, Mo and Zn), amino acids, vitamins, and plant growth regulators in them (Kumar et al. 2012). Furthermore, seaweeds and their derivatives are widely used in crop cultivation to enhance production systems. This is because various plant-growth-promoting substances (PGRs) such as auxin, cytokinin, and gibberellin have been identified as playing role in the fertility process. In this context, Mano and Nemoto (2012) documented that auxins are crucial modulators of plant development and growth. They support root growth, apical dominance, vascular development, flowering, and senescence. Auxins also play a role in tropic responses, cell division, elongation, and differentiation. Additionally, cytokinins are necessary for cell division and have an impact on a number of developmental processes, such as delaying senescence, breaking bud dormancy promoting, floral growth, maintaining apical dominance, mobilizing nutrient, promoting seed germination, and managing the root:shoot ratio (Sakakibara, 2006). Gibberellins also play a role in controlling floral and seed development, stem elongation, leaf expansion, and seed germination (Urbanová et al. 2013). Hashem et al. (2019) applied three species from distinct groups as soil amendments: *Ulva lactuca* (green algae), *Cystoseira* spp. (brown algae), and *Gelidium crinale* (red algae) in comparison to untreated plants. The purpose was to enhance the growth and yield of *Brassica napus* L. (canola) under normal conditions and salt stress conditions (NaCl 75 and 150 mM). Among the algae tested, *U. lactuca* exhibited significantly greater quantities of total carbohydrates, glycerol, proline, antioxidant activity, and phytohormones such as zeatin, benzyl adenine, and IAA (indole acetic acid), making it the most effective treatment. *G. crinale* had the highest levels of Indole butyric acid (AIB), while *Cystoseira* spp. had the highest levels of gibberellic acid (GA3) and jasmonic acid (JA). *Ascophyllum*, *Ecklonia*, *Fucus*, and *Macrocystis* are some of the most common genera of brown algae used in fertilizers and soil conditioners. Among them, *A. nodosum*, *F. serratus*, and *Laminaria digitata* are particularly favored in agriculture due to their size, high betaine cotent, and biomass. In addition, chemical fertilizers like urea and potash are not as effective as the red algae *Solieria robusta* in promoting the growth of soybean plants (*Glycine max* L.). These substances can protect plants from drying out, heat, and salinity (Haslam and Hopkins 1996). Franciraldo de Lima (2020). also provide sufficient levels of nitrogen and potassium but less phosphate compared to conventional animal fertilizers and chemical fertilizers with the usual N:P:K ratio. The alginic acid extracted from brown algae is used as a soil conditioner. It creates greater molecular weight polymers with metal ions that expand, especially in clay soils, retaining soil moisture, and improving soil structure. These cross-linked polymers improve the soil's aeration, capillary action in its pores, and water retention qualities (Gandhiyappan and Perumal, 2001).

3.2.4.2 Anti-phytopathogenic Activity of Macroalgae

Marine macroalgae can produce variety of effective biocidal compounds for managing plant diseases such as carbohydrates carotenoids, fatty acids, and polyphenols which are characterized by antimicrobial, and antiviral abilities toward many plant pathogens (Prisa et al. 2024). Seaweeds and their extracts also regulate many defense actions, including antioxidant enzymes, hormonal pathways, oxidative burst, and stress-related genes in plant (Bahmani et al. 2023). Carrageenans and laminarins have been shown to effectively activate defence responses in tobacco (*Nicotiana tabacum* L.) leaves (Mercier et al. 2001). The defense mechanism relies on identifying signal molecules called elicitors or plant systemic inducers that trigger a physiological response in the plant. Certain resistance inducers create molecules that mimic the elicitors or mimic the occurrence of an illness (Jayaraj et al. 2008). The application of seaweeds powders not only protect the crops from infection by root rotting fungi in naturally infested soil but also, they enhanced the plant growth such as vines length of watermelon, shoot lengths of eggplant and fresh shoot weights as compared to the control or the chemical fungicide "Topsin-M" (Baloch et al. 2013). The red seaweed "*Gracilaria confervoides*" can be considered a good alternative bioagent against pathogenic fungi such as *Fusarium solani*, *Rhizoctonia solani* and *Macrophomina phaseolina*" therefore; it may limit the wide usage of chemical fungicides in plant disease management which have side effects on the environment and human health (Soliman et al. 2018). Moreover, macroalgae have a high activity toward different plant viruses (plant cancer), whereas sulphate polysaccharides from brown seaweeds can block viral adsorption at the plant cell membrane (Jiao et al. 2011). Manzo et al. (2009) isolated betaines, dictyodial, dictyol C, and dicytol H from *Dictyota ciliolate*, and they demonstrated their antiviral activities against different plant viruses. The antiviral activity of kappa/beta-carrageenans extracted from red algae *Tichocarpus crinitus* against tobacco mosaic virus (TMV) in Xanthinc tobacco leaves (Nagorskaia et al. 2008).

Macroalgae is an important source of antinematodal bioactive compounds. *A. nodosum* and *Ecklonia maxima* extracts had a negative impact on eggs hatching and sensory perceptions of *Meloidogyne chitwoodi* and *M. hapla* on tomato plants (Ngala et al. 2016). Algal extracts increase plant resistance to nematodes, possibly by changing the auxin to cytokinin ratio. According to MacKinnon et al. (2010), betaines from *A. nodosum* also decreased infestations of *Meloidogyne javanica* and *M. incognita* in tomatoes. The nematode-suppressive impact of seaweeds in this investigation was similar to that of the chemical nematicide carbofuran. Furthermore, Soliera robusta, either alone or in conjunction with fertilizers or pesticides, significantly reduced nematode infection by preventing the development of galls on soybean roots (Haslam and Hopkins, 1996). Macroalgae are natural resources for the discovery of eco-friendly, environmental, and novel botanical insecticidal substances. Asharaja and Sahayaraj (2013) recorded that the chloroform extract of *S. swartzii* and *P. pavonica* caused nymphal mortality of *Dysdercus cingulatus* after 96 h. Moreover, the populations of red mite on strawberries were considerably decreased by biweekly spray treatments with a commercial product containing *A. nodosum* (Saifullah et al. 2007). Although the exact mechanism of action is unknown, the extracts may contain chelated metals that can lower mite fertility (Colavita et al. 2011). Additionally, the leaves' attractiveness to predatory insects may change as the amounts of phenolic and anthocyanin components increase. An estimated $100 billion is lost annually as a result of plant-parasitic nematodes (Zhang et al. 2006).

3.2.5 Green Biosynthesis of Nanoparticles Using Macroalgae

Nanotechnology is an integrated field of science and technology, which refers to the processing, separation, consolidation, and deformation of materials by atoms or molecules (Sharma et al. 2015). Nanobiotechnology is a cost-effective alternative to physical and chemical processes used in creating nanoparticles. The biosynthesis of nanoparticles (NPs) using green resources is a simple, environmentally friendly, pollutant-free, non-toxic, low-cost approach that requires less energy and time (Desai et al. 2021; El Newehy et al. 2025). Algal nanoparticles act as reducing and stabilizing agents, effectively reducing the negative impacts of radiation, heat, and electrical impulses from physical methods. They can also counteract the toxic effects of chemical substrates (Sharma et al. 2016). The ability of macroalgae to gather metals and decrease metal ions, as well as their capacity to produce NPs with a variety of sizes, forms, and morphologies, makes their use in biological NP synthesis intriguing (Desai et al. 2021). After one-step synthesis, the algal phytochemicals produce a strong coating on the metal nanoparticles by acting as an efficient metal-reducing and capping agent (Desai et al. 2021). The majority of phycosynthesis relies on an extracellular strategy, which uses bioactive chemicals from organic and aqueous algal extracts to synthesize nanoparticles (Desai et al. 2021; El Newehy et al. 2025).

Three main steps can be used to carry out the algae-mediated synthesis of nanoparticles, as explained below: (1) Heating or boiling algae for a set amount of time in water or an organic solvent to produce an algal extract. (2) The process involves creating molar solutions of ionic metallic compounds. (3) After that, incubating of algal solutions and molar solutions of ionic metallic compounds under controlled

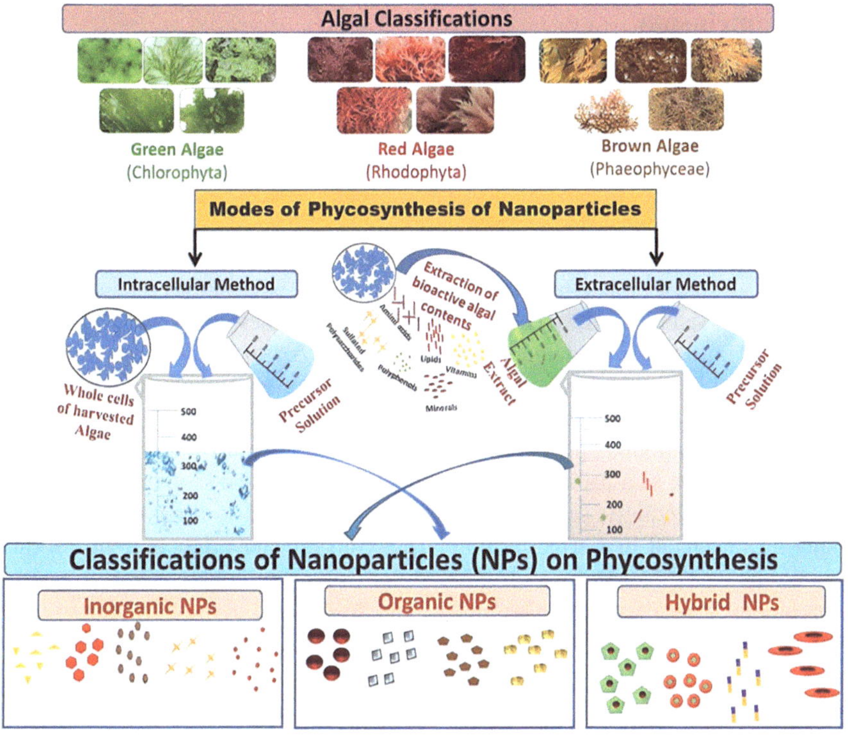

Fig. 3.9 Graphical representation of synthesis of various nanoparticles using three seaweeds classes (Desai et al. 2021)

conditions, either with constant stirring or without stirring for a specific period of time (Fig. 3.9).

In this connection, Ismail and Ismail (2016) documented that *C. prolifera*, *S. acinarium*, and *Gigartinia teedi* acted as bioreductants for the synthesis of nanostructured AgNPs ranging from 2 to 22 nm with antimicrobial activity against many human pathogenic microbes. Venkatraman et al. (2018) illustrated the antidiabetic, anti-inflammatory, and anticoagulant activity of Au-NPs from *T. conoides* that may be attributed to the presence of polyphenolic compounds, which act as a reducing agent in the formation of Au-NPs. Salem et al. (2019) recorded that iron oxide nanoparticles (Fe_3O_4-NPs) have been synthesized using brown (*C. sinuosa*) and red (*P. capillacea*) seaweed aqueous extracts, which are ideal candidates for pharmaceutical products and biomedical applications. Salem et al. (2020) confirmed the Fe_3O_4-NPs showed better antibiofilm efficiency than the tested seaweed. Recently, El Newehy et al. (2025) demonstrated that fucoidan-functionalized gold nanoparticles (F-AuNPs) synthesized from *S. cinereum* and *T. decurrens* exhibit promising antioxidant and anticancer properties.

3.3 Conclusion

Macroalgae are essential to the balance of the aquatic ecosystem and may offer a viable circular bioeconomy approach. They can remove toxins from wastewater and have been employed as a bioindicator in environmental conservation, particularly in monitoring water pollution. Seaweeds and their extracts have the potential to improve the quality of many foods by offering extra advantages. They can also be utilized to provide economic and nutritional benefits in aquaculture. Additionally, the bioactive substances found in seaweeds offer exciting opportunities for creating new natural drugs and cosmetics. The concept of carbon sequestration, which promotes the idea of mitigating climate change, can be realized by capturing carbon from seaweeds and converting it into biomethane through anaerobic digestion. Seaweeds have the potential to enhance human and animal health by reducing the reliance on chemical compounds in industrial processes. Additionally, it offers opportunities in agriculture as a packaging material, biostimulant, and biofuel producer. For the benefit of humanity, seaweeds are therefore predicted to bring about the next major advancement in marine farming. Thus, this chapter offers current information on the cost-effective uses of seaweed in a variety of fields, including agriculture, biofuel, pharmaceuticals, cosmetics, and health food. Reexamination of novel extraction techniques, safety concerns, and health-promoting qualities is necessary.

References

Abd Kadir MI, Wan Ahmad WY, Ahmad MR, Misnon MI et al (2014) Utilization of eco-colourant from green seaweed on textile dyeing. In: International colloquium in textile engineering, fashion, apparel and design, vol 6. Penang, Malaysia, pp 79–83

Abdel-Aziz MFA, Ragab MA (2017) Effect of use fresh macro algae (seaweed) *Ulva fasciata* and *Enteromorpha flaxusa* with or without artificial feed on growth performance and feed utilization of rabbit fish (*Siganus rivulatus*) fry. J Aquacult Res Dev 8:482

Abdul Khalil HPS, Lai TK, Tye YY, Rizal S, Chong EWN, Yap SW, Hamzah AA, Nurul Fazita MR, Paridah MT (2018) A review of extractions of seaweed hydrocolloids: properties and applications. eXPRESS Polym Lett 12(4):296–317

Abomohra A, El-Sheekh M, Hanelt D (2017) Screening of marine microalgae isolated from the hypersaline Bardawil lagoon for biodiesel feedstock. Renew Energy 101:266–1272

Abu Hafsa SH, Zahran SM, Ahmed MS, Abdel-Rahman N (2019) Effect of dietary supplementation with green and brown seaweeds on laying performance, egg quality, blood lipid profile and antioxidant capacity in Japanese quail. Poult Sci 39:41–59. https://doi.org/10.21608/epsj.2019.28828

Abudabos AM, Okab AB, Aljumaah RS, Samara EM et al (2013) Nutritional value of green seaweed (*Ulva lactuca*) for broiler chickens. Ital J Anim Sci 12:177–181

Abu-Ghannam N, Rajauria G (2013) Antimicrobial activity of compounds isolated from algae. In: Domínguez H (ed) Functional ingredients from algae for foods and nutraceuticals. Woodhead Publishing, pp 287–306. https://doi.org/10.1533/9780857098689.2.287

Adams Y, Smith SL, Schwartz-Albiez R, Andrews KT (2005) Carrageenans inhibit the in vitro growth of *Plasmodium falciparum* and cytoadhesion to CD36. Parasitol Res 97:290–294

Akbarzadeh S, Gholampour H, Farzadinia P, Daneshi A, Ramavandi B, Moazzeni A, Bargahi A (2018) Anti-diabetic effects of *Sargassum oligocystum* on Streptozotocin-induced diabetic rat. Iran J Basic Med Sci 21(3):342

Al-Harthi MA, El-Deek AA (2012) The effects of preparing methods and enzyme supplementation on the utilization of brown marine algae (*Sargassum dentifebium*) meal in the diet of laying hens. Ital J Anim Sci 10:48

Almela C, Clemente M, Velez D, Montoro R (2006) Total arsenic, inorganic arsenic, lead and cadmium contents in edible seaweed sold in Spain. Food Chem Toxicol 44:1901–1908

Al-Saif SSA, Abdel-Raouf N, El-Wazanani HA, Aref IA (2014) Antibacterial substances from marine algae isolated from Jeddah coast of Red sea, Saudi Arabia. Saudi J Biol Sci 21:57–64

Alzheimer's Association (2014) Alzheimer's disease facts and figures. Alzheimers Dement 10(2):e47–92. https://doi.org/10.1016/j.jalz.2014.02.001

Amster E, Tiwary A, Schenker M (2007) Case report: potential arsenic toxicosis secondary to herbal kelp supplement. Environ Health Perspect 115:606–608

Antizar-Ladislao B, Galil NI (2024) Biosorption of phenol and chlorophenols by acclimated residential biomass under bioremediation conditions in a sandy aquifer. Water Res 38:267–276

Arumugam N, Chelliapan S, Kamyab H, Thirugnana S, Othman N, Nasri NS (2018) Treatment of wastewater using seaweed: a review. Int J Environ Res Public Health 15:2851. https://doi.org/10.3390/ijerph15122851

Asharaja A, Sahayaraj K (2013) Screening of insecticidal activity of brown macroalgal extracts against *Dysdercus cingulatus* (Fab.) (Hemiptera: Pyrrhocoridae). J Biopest 6:193–203

Ayoola SO, Idowu AA (2008) Biotechnology and species development in aquaculture, pp 4722–4725

Bahmani R, More P, Babarinde S et al (2023) Seaweeds for plant disease management: current research advances and future perspectives. Phytoparasitica 51:783–802. https://doi.org/10.1007/s12600-023-01074-x

Bak HS (2010) A study on function of natural dyeing of seaweed. J Fish Mar Sci Educ 22(4):490–500

Baker CA, Martin AP, Yool A, Popova E (2022) Biological carbon pump sequestration efficiency in the north Atlantic: a leaky or a long-term sink? In: Global biogeochemical cycles, vol 36

Baloch GN, Tariq S, Ehteshamul-Haque S, Athar M et al (2013) Management of root diseases of eggplant and watermelon with the application of asafoetida and seaweeds. J Appl Bot Food Qual 86:138–142

Banach JL, Hoek-van den Hil EF, van der Fels-Klerx HL (2020) Food safety hazards in the European seaweed chain. Compr Rev Food Sci Food Saf 19:332–364

Barakat Kh, Ismail MM, Abou El Hassayeb H, El Sersy NA, Elshobary ME (2022) Chemical characterization and biological activities of ulvan extracted from *Ulva fasciata* (Chlorophyta). Rendiconti Lincei. Scienze Fisiche Naturali 33:829–841. https://doi.org/10.1007/s12210-022-01103-7

Barbot YN, Thomsen C, Thomsen L, Benz R (2015) Anaerobic digestion of *Laminaria japonica* waste from industrial production residues in laborarory- and pilot-scale. Mar Drugs 13:5947–5975

Battacharyya D, Babgohari MZ, Rathor P, Prithiviraj B (2015) Seaweed extracts as biostimulants in horticulture. Sci Hortic 196:39–48

Baweja P, Kumar S, Sahoo D, Levine I (2016) Biology of seaweeds. In: Seaweed in health and disease prevention. Academic Press, Cambridge, MA. https://doi.org/10.1016/B978-0-12-802772-1.00003-8

Bedoux G, Hardouin K, Burlot AS, Bourgougnon N (2014) Bioactive components from seaweeds: cosmetic applications and future development. Adv Bot Res 71:345–378

Bender DA (2003) Vitamin C (ascorbic acid). In: Nutritional biochemistry of the vitamins, 2nd edn. Cambridge University Press, UK

Berber I, Avsar C, Koyuncu H (2015) Antimicrobial and antioxidant activities of *Cystoseira crinita* Duby and *Ulva intestinalis* Linnaeus from the coastal region of Sinop, Turkey. J Coast Life Med 3:441–445

Berthon JY, Nachat-Kappes R, Bey M, Cadoret J-P, Renimel I, Filaire E (2017) Marine algae as attractive source to skin care. Free Radic Res 51(6):555–567. https://doi.org/10.1080/10715762.2017.1355550
Bhosale SH, Nagle VL, Jagtap TG (2002) Antifouling potential of some marine organisms from India against species of *Bacillus* and *Pseudomonas*. Mar Biotechnol 4:111–118
Bi L, Mou W, Sui H, Yang J, Yu J, Zhao L, Zou P (2018) Seaweed composition with oil-control function and cosmetic thereof. CN108904340A
Bird KT, Chynoweth DP, Jerger DE (1990) Effects of marine algal proximate composition on methane yields. J Appl Phycol 2:207–213
Birkemeyer C, Lemesheva V, Billig S, Tarakhovskaya E (2020) Composition of intracellular and cell wall-bound phlorotannin fractions in fucoid algae indicates specific functions of these metabolites dependent on the chemical structure. Metabolites 10:369. https://doi.org/10.3390/metabo10090369
Blunt JW, Copp BR, Munro MHG, Northcote PT, Prinsep MR (2011) Marine natural products. Nat Prod Rep 28:196–268
Bocanegra A, Bastida S, Benedi J et al (2009) Characteristics and nutritional and cardiovascular-health properties of seaweeds. J Med Food 12:236–258
Boettcher M, Chai F, Cullen J, Goeschl T, Lampitt R, Lenton A et al (2019) High level review of a wide range of proposed marine geoengineering techniques London: International Maritime Organization
Boyd PW, Bach LT, Hurd CL et al (2022) Potential negative effects of ocean afforestation on offshore ecosystems. Nat Ecol Evol 6:675–683. https://doi.org/10.1038/s41559-022-01722-1
Boyd KG, Mearns-Spragg AA, Brindley G, Hatzidimitrou K et al. (1998). Antifouling potential of epiphytic marine bacteria from the surfaces of algae. In: Le Gal Y, Muller-Feuga A (eds) Marine microorganisms for industry. IFREMER Plouzane, France, pp 128–136
Bradbury EJ, Wilkinson SJ, Cronin GM, Walk CL, Cowieson AJ (2012) The effect of marine calcium source on broiler leg integrity. In: Proceedings of 23rd Annual Australian poultry science symposium. Sydney, New South Wales, Australia, 19–22 February, pp 85–88
Canales-Gómez E, Correa G, Viana MT (2010) Effect of commercial carotene pigments (astaxanthin, cantaxanthin and β-carotene) in juvenile abalone *Haliotis rufescens* diets on the color of the shell or nacre. Vet Mexico Mexico 41:191–200
Cañedo-Castro B, Piñón-Gimate A, Carrillo S, Ramos D, Casas-Valdez M (2019) Prebiotic effect of *Ulva rigida* meal on the intestinal integrity and serum cholesterol and triglyceride content in broilers. J Appl Phycol 31:3265–3273. https://doi.org/10.1007/s10811-019-01785-x
Carlsson AS, Beilen JB, Moller R, Clayton D (2007) Micro-algae and macro-algae: utility for industrials applications (Bowles D, ed), pp 9–33
Cefalu WT (2001) Insulin resistance: cellular and clinical concepts. Exp Biol Med (Maywood) 226(1):13–26. https://doi.org/10.1177/153537020122600103. PMID: 11368233
Chan Y, Kim K, Cheah S (2011) Inhibitory effects of Sargassum polycystum on tyrosinase activity and melanin formation in B16F10 murine melanoma cells. J Ethnopharmacol 137:1183–1188.
Chandrasekaran M, Venkatesalu V, Adaikala Raj G (2014) Antibacterial activity of selected marine macro algae against vancomycin resistant Enterococcus faecalis. J Coast Life Med 2:940–946
Chandrasekharan N, Simmons DL (2004) The cyclooxygenases. Genome Biol 5:241. https://doi.org/10.1186/gb-2004-5-9-241
Chao P, Huang Y, Hsieh W (2013) Inhibitive effect of purple sweet potato leaf extract and its components on cell adhesion and inflammatory response in human aortic endothelial cells. Cell Adh Migr 7(2):237–245
Chapman VJ, Chapman DJ (1980) Seaweeds and their uses, 3rd edn. Chapman and Hall Ltd., London. https://doi.org/10.1007/978-94-009-5806-7
Chen Z, Zhang L, Liu Y, Wen Y, Shan S, Zhao C (2024) Seaweed as a sustainable future food source. Int J Food Sci Technol 59(3):1237–1247
Cheney D, Rajic L, Sly E, Meric D, Sheahan T (2014) Uptake of PCBs contained in marine sediments by the green macroalga *Ulva rigida*. Mar Poll Bull 88:207–214

Cheney D (2016) Toxic and harmful seaweeds. In: Fleurence J, Ira Levine I (eds) Seaweed in health and disease prevention. Academic Press, pp 407–421. https://doi.org/10.1016/B978-0-12-802772-1.00013-0

Chin YX, Lim PE, Maggs CA, Phang SM et al (2014) Anti-diabetic potential of selected Malaysian seaweeds. J Appl Phycol 27(5):1–12

Chirapart A, Ruangchuay R (2022) Seaweed in aquaculture: an overview. Sustain Global Resour Seaweeds 1:381–397. https://doi.org/10.1007/978-3-030-91955-9_20

Choi J-S, Moon WS, Choi JN, Do KH et al (2013) Effects of seaweed *Laminaria japonica* extracts on skin moisturizing activity in vivo. J Cosmet Sci 64:193–209

Chung IK, Sondak CF, Beardall J (2017) The future of seaweed aquaculture in a rapidly changing world. Eur J Phycol 52:495–505. https://doi.org/10.1080/09670262.2017.1359678

Chynoweth DP (2005) Renewable biomethane from land and ocean energy crops and organic wastes. HortScience 40:283–286

Chynoweth DP, Owens JM, Legrand R (2001) Renewable methane from anaerobic digestion of biomass. Renew Energy 22:1–8

Cirne-Santos CC, Barros CS, Nogueira CCR, Azevedo RC, Yamamoto KA, Meira GLS, de Vasconcelos ZFM, Ratcliffe NA, Teixeira VL, Schmidt-Chanasit J, Ferreira DF, Paixão ICNP (2019) Inhibition by marine algae of chikungunya virus isolated from patients in a recent disease outbreak in Rio de Janeiro. Front Microbiol 24(10):2426. https://doi.org/10.3389/fmicb.2019.02426

Colavita RV, Anderson SM, Ford GW, Robinson JS, Wideman RF Jr (2011) A study of certain environmental factors and mineral chelation on the reproductive performance of young and yearling Turkey hens. Poult Sci 90:303–308

Cotas J, Gomes L, Pacheco D, Pereira L (2023) Ecosystem services provided by seaweeds. Hydrobiology 2(1):75–96. https://doi.org/10.3390/hydrobiology2010006

Cruywagen CW, Taylor S, Beya MM, Calitz T (2015) The effect of buffering dairy cow diets with limestone, calcareous marine algae, or sodium bicarbonate on ruminal pH profiles, production responses, and rumen fermentation. J Dairy Sci 98:5506–5514. https://doi.org/10.3168/jds.2014-8875

Cruz-Suárez LE, Tapia-Salazar M, Nieto-López MG, Ricque-Marie D (2009) Use of seaweeds for shrimp nutrition: status and potential. In: Browdy CL, Jory DE (eds) The rising tide. Proceedings of the special session on sustainable shrimp farming. World Aquaculture Society, Baton Rouge, pp 147–163

Cyrus MD, Bolton JJ, Scholtz R, Macey BM (2014) The advantages of *Ulva* (Chlorophyta) as an additive in sea urchin formulated feeds: effects on palatability, consumption and digestibility. Aquacult Nutr 1–14

da Gama BAP, Ana GV, Weidner K, Soares AR et al (2008) Antifouling activity of natural products from Brazilian seaweeds. Bot Mar 51:191–201

Dahms HU, Dobretsov S (2017) Antifouling compounds from marine macroalgae. Mar Drugs 15:265

Das S, Behera M, Behera PR, Adarsh V, Singh L (2025) Seaweed biosorption: a sustainable approach to wastewater treatment. In: Dey S, Shah MP (eds) Biotechnological removal of emerging pollutants from wastewater systems. Advances in wastewater research. Springer, Singapore. https://doi.org/10.1007/978-981-96-3945-8_6

Desai N, Pawar U, Aparadh V, Dethe U, Gaikwad D (2021) Seaweeds: a potential source in progressing nanotechnology. In: Thangadurai D, Sangeetha J, Prasad R (eds) Bioprospecting algae for nanosized materials. Springer Nature Switzerland AG. https://doi.org/10.1007/978-3-030-81557-8

Devaki T, Sathivel A, Balaji HR (2009) Stabilization of mitochondrial and microsomal function by polysaccharide of *Ulva lactuca* on D-Galactosamine induced hepatitis in rats. Chem Biol Interact 177:83–88

Deviller G (2003) Traitement par lagunage à haut rendement algal (LHRA) des effluents piscicoles marins recyclés: évaluation chimique et écotoxicologique. Thèse d'Université, Montpellier I

Dhargalkar VK, Verlecar XN (2009) Southern ocean seaweeds: a resource for exploration in food and drugs. Aquacult 287:229–242. https://doi.org/10.1016/j.aquaculture.2008.11.013

Di Liegro I (2008) Thyroid hormones and the central nervous system of mammals (Review). Mol Med Rep 1(3):279–295

Díaz AC, Espino ML, Arzoz NS, Velurtas SM et al (2017) Fenucci free radical scavenging activity of extracts from seaweeds *Macrocystis pyrifera* and *Undaria pinnatifida*: applications as functional food in the diet of prawn *Artemesia longinaris* seaweed extracts in diets for *Artemesia longinaris*. Lat Am J Aquat Res 45(1):104–112

Dinakarkumar Y, Ramakrishnan G, Mekala JR, Reddy MM, John RM, Raj ALT, Chamarthy S (2025) Exploration of marine macro algal-derived compounds as antiviral drugs: a review on the types, mechanisms of action, and challenges. Pharmacol Res Nat Prod. https://www.sciencedirect.com/science/article/pii/S2950199725000758

Ding T, Yang M, Zhang J, Yang B, Lin K, Juying Li J, Gan J (2017) Toxicity, degradation and metabolic fate of ibuprofen on freshwater diatom *Navicula* sp. J Hazard Mater 330:127–134. https://doi.org/10.1016/j.jhazmat.2017.02.004

Dokulil MT (2003) Algae as ecological bio-indicators. In: Markert BA, Breure AM, Zechmeister HG (eds) Bioindicators and biomonitors. Elsevier, Amsterdam, pp 285–327

Dolui AK (2021) Marine bioprospecting: seaweeds for industrial molecules. Bioprospect Plant Biodivers Ind Mol. https://doi.org/10.1002/9781119718017.ch19

Doty S, Aguilar-Santos G (1966) Caulerpicin, a toxic constituent of caulerpa. Nature 11:990. https://doi.org/10.1038/211990a0

Duarte CM, Wu J, Xiao X, Bruhn A, Krause-Jensen D (2017) Can seaweed farming play a role in climate change mitigation and adaptation? Front Mar Sci 4:100

Duarte CM, Bruhn A, Krause-Jensen D (2021) A seaweed aquaculture imperative to meet global sustainability targets. Nat Sustain 1–9. https://doi.org/10.1038/s41893-021-00773-9

El Newehy AS, Gheda SF, Ismail MM, Aldisi D, Abulmeaty MMA, Elshobary ME (2025) Fucoidan-based gold nanoparticles; antioxidant and anti-cancer potential from *Turbinaria decurrens* and *Sargassum cinereum*. Pharmaceutics 17:826. https://doi.org/10.3390/pharmaceutics17070826

El Zokm GM, Ibrahim MIA, Mohamed LA, El-Mamoney M (2020) Critical geochemical insight into Alexandria coast with special reference to diagnostic ratios (TOC/TN & Sr/Ca) and heavy metals ecotoxicological hazards. Egypt J Aquat Res 46:27–33

El Zokm GM, Ismail MM, El-Said GhF (2021) Halogen content relative to the chemical and biochemical composition of fifteen marine macro and micro algae: nutritional value, energy supply, antioxidant potency, and health risk assessment. Environ Sci Pollut Res 28(12):14893–14908. https://doi.org/10.1007/s11356-020-11596-0

El Zokm GM, Ismail MM, El Okba M (2022) Seaweed as bioindicator of organic micropollutants polycyclic aromatic hydrocarbons (PAHs) and organochlorine pesticides (OCPs). Environ Sci Pollut Res. https://doi.org/10.1007/s11356-022-18634-z

El-Maaty HAA, EL-Khateeb AY, Al-Khalaifah H, El Hamed EA, Hamed S, El-Said EA, Mahrose KM, Metwally K, Mansour AM (2021) Effects of ecofriendly synthesized calcium nanoparticles with biocompatible *Sargassum latifolium* algae extract supplementation on egg quality and scanning electron microscopy images of the eggshell of aged laying hens. Poult Sci 100:675–684. https://doi.org/10.1016/j.psj.2020.10.043

Elmaidomy AH, Zahran EM, Soltane R, Alasiri A, Saber H, Ngwa CJ, Pradel G, Alsenani F, Sayed AM, Abdelmohsen UR (2022) New halogenated compounds from *Halimeda macroloba* seaweed with potential inhibitory activity against Malaria. Molecules 27:5617. https://doi.org/10.3390/molecules27175617

El-Said GhF, El Zokm GM, El-Sikaily A, Ismail MM (2024) Comparison of crude nano-hydroxyapatite extracted from calcified seaweed in terms of composition, antibacterial activity, and cytotoxicity. Environ Nanotechnol Monit Manage 21:100908. https://doi.org/10.1016/j.enmm.2023.100908

El-Sheekh MM, Ismail MM, Hamouda MM (2016) Influence of some brown seaweed extracts on germination and cytological responses of *Trigonella foenum-graecum* L. BioTechnol Indian J 12(9):104–116

Evans FD, Critchley AT (2014) Seaweeds for animal production use. J Appl Phycol 26:891–899

Farghali M, Mohamed IMA, Osman AI, Rooney DW (2023) Seaweed for climate mitigation, wastewater treatment, bioenergy, bioplastic, biochar, food, pharmaceuticals, and cosmetics. Review 21:97–152

Fernand F, Israel A, Skjermo J, Wichard T, Timmermans KR, Golberg A (2017) Offshore macroalgae biomass for bioenergy production: Environmental aspects, technological achievements and challenges. Renew Sustain Energy Rev 75:35–45. https://doi.org/10.1016/j.rser.2016.10.046

Fitzgerald C, Gallagher E, Tasdemir D et al (2011) Heart health peptides from macroalgae and their potential use in functional foods. J Agric Food Chem 59:6829–6836

Flannery T (2017) Sunlight and seaweed: an argument for how to feed, power and clean up the world. Text Publishing, Melbourne

Francavilla M, Franchi M, Monteleone M et al (2013) The red seaweed Gracilaria gracilis as a multi products source. Mar Drugs 11(10):3754–3776

Franciraldo de Lima J (2020) Utilization of *Chlorella* spp. as biostimulant in the germination of melon seeds (*Cucumis melo* L.). J Agric Stud 8(2):750

Froehlich HE, Afflerbach JC, Frazier M, Halpern BS (2019) Blue growth potential to mitigate climate change through seaweed offsetting. Curr Biol 29:3087–3093(e3083). https://doi.org/10.1016/j.cub.2019.07.041

Gandhiyappan K, Perumal P (2001) Growth promoting effect of seaweed liquid fertilizer (*Enteromorpha intestinalis*) on the sesame crop plant. Seaweed Res Util 23:23–25

Garcia AN, Stein EM, Leonardo Zambotti Villela LZ, Yokoya NS, Neto PC, de Carvalho LR (2020) *Dichotomaria marginata* (Rhodophyta) as a bioindicator for marine pollution: an overview about its metabolites and adsorbed pollutants Revista Biología Marina y Oceanografía 55(2):128–141. https://doi.org/10.22370/rbmo.2020.55.2.2498

García-Bueno N, Decottigniesa P, Turpin VT, Fleurence J et al (2014) Seasonal antibacterial activity of two red seaweeds, *Palmaria palmata* and *Grateloupia turuturu*, on European abalone pathogen *Vibrio harveyi*. Aquat Living Resour 27(27):83–89

Germain S, Bum EN, Talla E, Dimo T et al (2011) Antipyretic and antinociceptive effects of *Nauclealatifolia* roots decoction and possible mechanisms of action. Pharm Biol 49(1):15–25

Gómez-Guzmán M, Rodríguez-Nogales A, Algieri F, Potential JG (2018) Role of seaweed polyphenols in cardiovascular-associated disorders. Mar Drugs 16(25):1–21

Guillerme J-B, Couteau C, Coiffard L (2017) Applications for marine resources in cosmetics. Cosmetics 4:35

Gunerken E, D'Hondt E, Eppink MH, Garcia-Gonzalez L et al (2015) Cell disruption for microalgae biorefneries. Biotechnol Adv 33:243–260

Gurung A, Van Ginkel SW, Kang W, Qambrani N, Oh S (2012) Evaluation of marine biomass as a source of methane in batch tests: a lab-scale study. Energy 43:396–401

Hallsson SV (1964) The uses of seaweed in Iceland. In: Proceedings of the Fourth International seaweed symposium, Biarritz, France. Macmillan, New York. pp 398–405

Han JS, Lee EJ, Lee JW, Oh JY, Park SG (2017) Sunscreen cosmetic composition. KR20170004842A

Hanssen JF, Indergaard M, Østgaard K, Bævre OA, Pedersen TA, Jensen A (1989) Biogas production from Laminaria hyperborea, Laminaria saccharina and Ascophyllum nodosum in semicontinuous cultures. Biomass 14:1–13

Hansson G (1983) Methane production from marine, green macro-algae. Resour Conservation. 8:185–194

Harder T, Dobretsov S, Qian PY (2004) Waterborne polar macromolecules act as algal antifoulants in the seaweed Ulva reticulata. Mar Ecol Prog Ser 274:133–141

Harichandan S (2025) Harnessing seaweed for biohydrogen production: an integrated ecological and energy framework. Int J Hydrogen Energy 136:1–10. https://doi.org/10.1016/j.ijhydene.2025.05.067

Hashem H, Mansour H, El-Khawas S, Hassanein R (2019) The potentiality of marine macro-algae as biofertilizers to improve the productivity and salt stress tolerance of canola (Brassica napus L.). Plants Agron 9:146

Haslam SFI, Hopkins DW (1996) Physical and biological effects of kelp (seaweed) added to soil. Appl Soil Ecol 3:257–261

Hasna AF, Junianto (2021) Utilization of seaweed industry solid waste for paper raw materials. Glob Sci J 9(11):681–685

Hassan Q, Algburi S, Sameen AZ, Salman HM, Jaszczur M (2024) Green hydrogen: a pathway to a sustainable energy future. Int J Hydrogen Energy 50:310–333. https://doi.org/10.1016/J.IJHYDENE.2023.08.321

Hatcher H, Planalp R, Cho J, Torti FM, Torti SV (2008) Curcumin: from ancient medicine to current clinical trials. CMLS 65(11):1631–1652

He P, Xu S, Zhang H, Wen S, Dai Y, Lin S et al (2008) Bioremediation efficiency in the removal of dissolved inorganic nutrients by the red seaweed, porphyra yezoensis, cultivated in the open sea. Water Res 42:1281–1289. https://doi.org/10.1016/j.watres.2007.09.023

Hell A, Labrie S, Beaulieu L (2018) Effect of seaweed flakes addition on the development of bioactivities in functional Camembert-type cheese. Int J Food Sci Technol 53:1054–1064. https://doi.org/10.1111/ijfs.13681

Heo SJ, Jeon YJ (2009) Protective effect of fucoxanthin isolated from Sargassum siliquastrum on UV-B induced cell damage. J Photochem Photobiol B 95:101–107. https://doi.org/10.1016/j.jphotobiol.2008.11.011

Hernández-Carmona G, Freile-Pelegrín Y, Hernández-Garibay E (2013) Conventional and alternative technologies for the extraction of algal polysaccharides. In: Domnínguez H (ed) Functional ingredients from algae for foods and nutraceuticals. Woodhead Publishing Limited, Cambridge, UK, pp 475–516

Hofmann D, Singh D, Ebonhoh O (2018) Evaluating potential of green alga Chlorella Vulgaris to accumulate phosphorus and to fertilize nutrient-poor soil substrates for crop plants. J Appl Phycol 30:2827–2836

Holdt SL, Kraan S (2011) Bioactive compounds in seaweed: functional food applications and legislation. J Appl Phycol 23:543–597

Holt S (2008) Fucose complexes, fucoxanthin, fucoid and fat storage. Townsend Lett 87–92

Holtkamp AD, Kelly S, Ulber R, Lang S (2009) Fucoidans and fucoidanas esfocus on techniques for molecular structure elucidation and modification of marine polysaccharides. Appl Microbiol Biotechnol 82(1):1–11

Hong D, Hoang M-H, Lan Anh H (2011) Studies on the analgesic and anti-inflammatory activities of *Sargassum swartzii* (Turner) C. Agardh (Phaeophyta) and *Ulva reticulata* Forsskal (Chlorophyta) in experiment animal models. Afr J Biotechnol 10:2308–2314

Horn MH, Messer KH (1992) Fish guts as chemical reactor: a model of the alimentary canals of marine herbivorous fishes. Biology 113:527–535

Hossain A, Knorr G, Jokat W, Lohmann G (2021) Climate model results of Fram Strait widening sensitivity studies of COSMOS in NetCDF, p 19

Hotchkiss S (2011) Seaweed rediscovered: healthy hydrocolloids and beyond, cybercolloids. Strand Farm, Currabinny, Carrigaline, Co., Cork, Ireland)

Huang J, Li C, Lin S, Liu D, Ma J, Qian J, Xiang Q (2019) Water-free washing-free gold seaweed cleaning oil. CN109431878A

Huang HL, Wang BG (2004) Antioxidant capacity and lipophilic content of seaweeds collected from the Qingdao coastline. J Agric Food Chem 52:4993–4997

Huesemann M, Roesjadi G, Benemann J, Metting FB (2010) Biofuels from microalgae and seaweeds. In: Biomass to biofuels: strategies for global industries. Wiley, West Sussex, UK

Hyun JH, Kim SC, Kang JI, Kim MK et al (2009) Apoptosis inducing activity of fucoidan in HCT-15 colon carcinoma cells. Biol Pharm Bull 32:1760–1764

Inouhe M (2005) Phytochelatins. Braz J Plant Physiol 17:65–78

Irfan M, Wahab IH, Sarni, Subur R, Akbar N (2019) Seaweed *Sargassum* sp. as material for biogas production. AACL Bioflux 12(5):2015–2019

Ismail MM (2022) Probiotics and algal functional food. In: Mishra A (ed) Algal functional foods and nutraceuticals: benefits, opportunities, and challenges. Bentham Science Publishers, Sharjah, pp 341–361

Ismail MM, Amer MS (2020) Characterization and biological properties of sulfated polysaccharides of *Corallina officinalis* and *Pterocladia capillacea*. Acta Botanica Brasilica 34(4):623–632. https://doi.org/10.1590/0102-33062020abb0121

Ismail MM, El Zokm GM (2025) Algae as keystone for blue economy: sustainability and challenges "Review." Discov Sustain 6:366. https://doi.org/10.1007/s43621-024-00746-w

Ismail MM, El-Shafay SM (2015) Variation in taxonomical position and biofertilizing efficiency of some seaweed on germination of *Vigna unguiculata* (L). Int J Environ Sci Eng 6:47–57

Ismail GA, Ismail MM (2016) Comparative study of silver nanoparticles green biosynthesis using different seaweeds and their antimicrobial activity against some human pathogens. Ciencia Tecnica Vitivinicola J 21(2):29–55

Ismail GA, Ismail MM (2017) Variation in oxidative stress indices of two green seaweeds growing under different heavy metal stresses. Environ Monit Assess 189:68–80

Ismail MM, Gheda SF, Pereira L (2016) Variation in bioactive compounds in some seaweeds from Abo Qir bay, Alexandria, Egypt. Rendiconti Lincei di Scienze Fisiche e Naturali Journal 27(2):269–279

Ismail MM, El Zokm GM, El-Sayed AM (2017) Variation in biochemical constituents and master elements in common seaweeds from Alexandria Coast, Egypt, with special reference to their antioxidant activity and potential food uses: prospective equations. Environ Monit Assess 189:648

Ismail MM, El Zokm GM, Lopez JM (2023b) Nutritional, bioactive compounds content, and antioxidant activity of brown seaweeds from the Red Sea. Front Nutr 10:1210934. https://doi.org/10.3389/fnut.2023

Ismail MM, Ibrahim HAH, El Zokm GM (2024) Comparison of chemical properties and bioactivities of different polysaccharides types extracted from brown seaweeds habited in the Egyptian Red Sea, Hurghada. Hydrobiol J 60(3):56–79

Ismail MM, Kanaan HM (2022) Chemodiversity of galactans from different red seaweed species with emphasis on their antioxidant and antiproliferative activities in vitro. Braz Arch Biol Technol 65:e22210122. https://doi.org/10.1590/1678-4324-2022210122

Ismail MM, Alotaibi BS, EL-Sheekh MM (2020a) Therapeutic uses of red macroalgae "Review". Molecules 25:4411. https://doi.org/10.3390/molecules25194411

Ismail MM, Ismail GA, El-Sheekh· MM (2020b) Potential assessment of some micro- and macroalgal species for bioethanol and biodiesel production. Energy Sources, Part A: Recov Util Environ Effects 1–17. https://doi.org/10.1080/15567036.2020.1758853

Ismail MM, Mostafa MH, El Sikaily AM (2022) Phytochemical variation, antioxidant and antidiabetic capacity of extracts of common brown seaweeds from the Red Sea, Egypt. Egypt J Aquat Biol Fish 26(5):445–462. https://doi.org/10.21608/EJABF.2022.262605

Ismail MM, El Zokm G, El Sikaily AM, Selim AI, Ismail GA (2023a) Chemodiversity and bioactivity assessment of phlorotannins from some phaeophyta species from the Red Sea. J Appl Phycol 35:1769–1788. https://doi.org/10.1007/s10811-023-03000-4

Ismail MM, El-Fakharany EM, Hegazy GhE (2023c) Purification and fractionation of phycobiliproteins from *Arthrospira platensis* and *Corallina officinalis* with evaluating their biological activities. Sci Rep 13:14270. https://doi.org/10.1038/s41598-023-41001-y

Ismail MM, Lopez JMM, El-Sayed AAM (2025) Evaluation of the spatial variations in the biochemical composition of seaweed species along the coast of Alexandria, with a focus on fatty acids

and total amino acids of the prevalent edible species. Food Sci Nutr 1–19. https://doi.org/10.1002/fsn3.70645
Jacobs RD, Gordon MBE, Vineyard KR, Keowen ML, Garza F Jr, Andrews FM (2020) The effect of a seaweed-derived calcium supplement on gastric juice pH in the horse. J Equine Vet Sci 95:103265. https://doi.org/10.1016/j.jevs.2020.103265
Jagtap AS, Meena SN (2022) Seaweed farming: a perspective of sustainable agriculture and socio-economic development. In: Jhariya MK, Meena RS, Banerjee A, Meena SN (eds) Natural resources conservation and advances for sustainability. Elsevier, Amsterdam, pp 493–501. https://doi.org/10.1016/b978-0-12-822976-7.00022-3
Jayaraj J, Wan A, Rahman M, Punja ZK (2008) Seaweed extract reduces foliar fungal diseases on carrot. Crop Prot 27(10):1360–1366
Jeevithan L, Elango J, Sanchez De Val JE, Wu W (2025) Marine-derived novel compounds for fibrinolysis. Biochemistry. IntechOpen. https://doi.org/10.5772/intechopen.115095
Jesumani V, Du H, Aslam M, Pei P, Huang N (2019) Potential use of seaweed bioactive compounds in skincare a review. Mar Drugs 17:688. https://doi.org/10.3390/md17120688
Jhariya MK et al (2021) Ecological intensification of natural resources towards sustainable productive system. In: Ecological intensification of natural resources for sustainable agriculture. Springer, pp 1–28. https://doi.org/10.1007/978-981-33-4203-3_1
Jiao G, Yu G, Zhang J, Ewart HS (2011) Chemical structures and bioactivities of sulfated polysaccharides from marine algae. Mar Drugs 9:196–223
Joly AB (1967) Gêneros de algas marinhas da costa atlântica latino-americana. São Paulo: Ed. da Universidade, p 461
Kageyama H, Waditee-Sirisattha R (2019) Antioxidative, anti-inflammatory, and anti-aging properties of mycosporine-like amino acids: molecular and cellular mechanisms in the protection of skin-aging. Mar Drugs 17:222. https://doi.org/10.3390/md17040222
Kaliaperumal K, Aboobacker J, Dhanapal V, Gayathri A, Subramanian K, Suresh K et al (2025) Bioremediation of malachite green dye using *Sargassum wightii* seaweed and its biological and physicochemical characterization. Open Chem 23(1). https://doi.org/10.1515/chem-2025-0137
Kanagasabhapathy M, Yamazaki G, Ishida A, Sasaki H, Nagata S (2009) Presence of quorum-sensing inhibitor-like compounds from bacteria isolated from the brown alga *Colpomenia sinuosa*. Lett Appl Microbiol 49:573–579
Khan N, Sudhakar K, Mamat R (2025) Bio-based papers from seaweed and coconut fiber: sustainable materials for a greener future. Carbon Resour Convers 100329
Khursheed M, Ghelani H, Jan RK, Adrian TE (2023) Anti-inflammatory effects of bioactive compounds from seaweeds, bryozoans, jellyfish, shellfish and peanut worms. Mar Drugs 21(10):524. https://doi.org/10.3390/md21100524
Kim CR, Kim YM, Lee MK, Kim IH et al (2017) Pyropia yezoensis peptide promotes collagen synthesis by activating the TGF-b/Smad signaling pathway in the human dermal fibroblast cell line Hs27. Int J Mol Med 39:31–38
Kim E, Cui J, Kang I, Zhang G, Lee Y (2021) Potential antidiabetic effects of seaweed extracts by upregulating glucose utilization and alleviating inflammation in C2C12 myotubes. Int J Environ Res Public Health 2021(18):1367
Kishimoto M, Okakura T, Nagashima H, Minowa T et al (1994) CO_2 fixation and oil production using microalgae. J Ferment Bioeng 78:479–482
Kjelleberg S, Steinberg M, Givskov L, Gram M et al (1997) Do marine natural products interfere with prokaryotic AHL regulatory systems? Aquat Microbiol Ecol 13:85–93
Kotelnikova S (2002) Microbial production and oxidation of methane in deep subsurface. Earth-Sci Rev 58:367–395. https://doi.org/10.1016/S0012-8252(01)00082-4
Krause-Jensen D, Duarte CM (2016) Substantial role of macroalgae in marine carbon sequestration. Nat Geosci 9:737–742. https://doi.org/10.3389/fmars.2019.00375
Kulshreshtha G, Rathgeber B, Stratton G, Thomas N, Evans F, Critchley A, Hafting J, Prithiviraj B (2014) Feed supplementation with red seaweeds, *Chondrus crispus* and *Sarcodiotheca*

gaudichaudii, affects performance, egg quality, and gut microbiota of layer hens. Poult Sci 93:2991–3001

Kulshreshtha G, Rathgeber B, MacIsaac J, Boulianne M, Brigitte L, Stratton G, Thomas NA, Critchley AT, Hafting J, Prithiviraj B (2017) Feed supplementation with red seaweeds, *Chondrus crispus* and *Sarcodiotheca gaudichaudii*, reduce *Salmonella enteritidis* in laying hens. Front Microbiol 8:567. https://doi.org/10.3389/fmicb.2017.00567

Kumar MS, Sharma SA (2021) Toxicological effects of marine seaweeds: a cautious insight for human consumption. Crit Rev Food Sci Nutr 61(3):500–521. https://doi.org/10.1080/10408398.2020.1738334. Epub 2020 Mar 19 PMID: 32188262

Kumar V, Gupta P, Dwivedi S (2012) Bio-efficacy of fly-ash based trichoderma formulations against damping-off and root-rot diseases in tomato (*Lycopersicum esculentum*). Indian Phytopathol 65(4):404–405

Kuznetsova TA, Andryukov BG, Makarenkova ID, Zaporozhets TS, Besednova NN, Fedyanina LN, Kryzhanovsky SP, Shchelkanov MY (2021) The potency of seaweed sulfated polysaccharides for the correction of hemostasis disorders in COVID-19. Molecules 26:2618. https://doi.org/10.3390/molecules2609261

Lahaye M, Kaeffer B (1997) Seaweed dietary fibres: structure, physico-chemical and biological properties relevant to intestinal physiology. Sci Aliment 17:563–584

Larson S, Stoeckl N, Fachry ME, Mustafa MD, Lapong I, Purnomo AH et al (2021) Women's well-being and household benefits from seaweed farming in Indonesia. Aquaculture 530:735711. https://doi.org/10.1016/j.aquaculture.2020.735711

Laurens L, Chen-Glasser M (2017) A perspective on renewable bioenergy from photosynthetic algae as feedstock for biofuels and bioproducts. Elsevier

Lee HJ, Kim HC, Vitek L, Nam CM (2010) Algae consumption and risk of type 2 diabetes: Korean National Health and Nutrition Examination Survey in 2005. J Nutr Sci Vitaminol (Tokyo) 56:13–18

Lee SH, Kang SM, Ko SC, Kang MC, Jeon YJ (2013) Octaphlorethol a, a novel phenolic compound isolated from Ishige foliacea, protects against streptozotocin-induced pancreatic beta cell damage by reducing oxidative stress and apoptosis. Food Chem Toxicol 59:643–649

Leslie SK, Cohen DJ, Sedlaczek J, Pinsker EJ et al (2013) Controlled release of rat adipose-derived stem cells from alginate microbeads. Biomaterials 34:8172–8184

Li Q, Luo J, Wang C, Tai W, Wang H, Zhang X, Liu K, Jia Y, Lyv X, Wang L, He H (2018) Ulvan extracted from green seaweeds as new natural additives in diets for laying hens. J Appl Phycol 30(3):2017–2027. https://doi.org/10.1007/s10811-017-1365-2

Lindenmayer R, Lu L, Eivazi F, Afrasiabi Z (2023) Atomic spectroscopy-based analysis of heavy metals in seaweed species. Appl Sci13:4764. https://doi.org/10.3390/app13084764

Liu H, Wang G (2014) Fermentative hydrogen production from macroalgae *Laminaria japonica* using anaerobic mixed bacteria. Int J Hydrogen Energy 39(16):9012–9017. https://doi.org/10.1016/j.ijhydene.2014.03.244

Lomartire S, Gonçalves AMM (2022) Antiviral activity and mechanisms of seaweeds bioactive compounds on enveloped viruses-a review. Mar Drugs 20(6):385. https://doi.org/10.3390/md20060385. PMID: 35736188; PMCID: PMC9228758

Lomartire S, Marques JC, Gonçalves AMM (2021) An overview to the health benefits of seaweeds consumption. Mar Drugs. https://doi.org/10.3390/md19060341

López-Hortas L, Flórez-Fernández N, Torres MD, Ferreira-Anta T, Casas MP, Balboa, Lymperatou A, Engelsen TK, Skiadas IV, Gavala HN (2022) Different pretreatments of beach-cast seaweed for biogas production. J Clean Prod 362. https://doi.org/10.1016/j.jclepro.2022.132277

Lymperatou A, Engelsen T, Skiadas IV, Gavala HN (2022) Different pretreatments of beach-cast seaweed for biogas production. J Clean Prod 362:132277. https://doi.org/10.1016/j.jclepro.2022.132277

Lyons NM, O'Brien NM (2002) Modulatory effects of an algal extract containing astaxanthin on UVA-irradiated cells in culture. J Dermatol Sci 30:73–84

Machado LP, Carvalho LR, Young MCM, Cardoso-Lopes E-M et al (2015) Evaluation of acetylcholinesterase inhibitory activity of Brazilian red macroalgae organic extracts. Rev Bras 25(6):657–662
MacKinnon SL, Hiltz D, Ugarte R, Craft CA (2010) Improved methods of analysis for betaines in *Ascophyllum nodosum* and its commercial seaweed extracts. J Appl Phycol 22:489–494
Macreadie PI, Anton A, Raven JA, Beaumont N, Connolly RM, Friess DA et al (2019) The future of blue carbon science. Nat Commun 10:1–13. https://doi.org/10.1038/s41467-019-11693-w
Maeda H, Hosokawa M, Sashima T, Murakami-Funayama K, Miyashita K (2009) Anti-obesity and anti-diabetic effects of fucoxanthin on diet-induced obesity conditions in a murine model. Mol Med Rep 2:897–902
Mahapatra MK, Kumar A (2017) A short review on biobutanol, a second-generation biofuel production from lignocellulosic biomass. J Clean Energy Technol 5:27–30
Mahrose KM, Michalak I (2022) Seaweeds for animal feed, current status, challenges, and opportunities. In: Ranga Rao A, Ravishankar GA (eds) Sustainable global resources of seaweeds, vol 1. Springer, Cham, pp 357–379. https://doi.org/10.1007/978-3-030-91955-9_19
Makkar HPS, Tran G, Heuzé V, Giger-Reverdin S et al (2016) Seaweeds for livestock diets: a review. Anim Feed Sci Technol 212:1–17
Manikandan DB, Veeran S, Seenivasan S et al (2022) Exploration of marine red seaweed as a dietary fish meal replacement and its potentiality on growth, hematological, biochemical, and enzyme activity in freshwater fish Labeo rohita. Trop Anim Health Prod 54:395. https://doi.org/10.1007/s11250-022-03392-4
Mano Y, Nemoto K (2012) The pathway of auxin biosynthesis in plants. J Exp Bot 63:2853–2872
Manzo E, Ciavatta ML, Bakkas S, Villani G et al (2009) Diterpene content of the alga *Dictyota ciliolata* from a Moroccan lagoon. Phytochem Lett 2:211–215
Maroli L, Pavoni B, Sfriso A, Raccanelli S (1993) Concentrations of polychlorinated biphenyls and pesticides in different species of macroalgae from the Venice lagoon. Mar Pollut Bull 26:553–558
Matanjun P, Mohamed SM, Kharidah M, Noordin MM (2010) Comparison of cardiovascular protective effects of tropical seaweeds, *Kappaphycus alvarezii*, *Caulerpa lentillifera*, and *Sargassum polycystum*, on highcholesterol/ high-fat diet in rats. J Med Food 13:792–800
McHugh DJ (2003) A guide to the seaweed industry. FAO Fisheries Technical Papers 441, p 105
Mendis E, Kim SK (2011) Present and future prospects of seaweeds in developing functional foods. Adv Food Nutr Res 64:1–15
Mercier L, Lafitte C, Borderies G, Briand X, Esquerre-Tugaye MT, Fournier J (2001) The algal polysaccharide carrageenans can act as an elicitor of plant defence. New Phytol 149:43–51
Mhadhebi L, Mhadhebi A, Robert J, Bouraoui A (2014) Antioxidant, anti-inflammatory and antiproliferative effects of aqueous extracts of three mediterranean brown seaweeds of the genus Cystoseira. Iran J Pharm Res. 13(1):207–220
Milledge JJ, Nielsen BV, Maneein S, Harvey PJ (2019) A brief review of anaerobic digestion of algae for bioenergy. Energies 12:1–22
Miyashita K, Nishikawa S, Beppu F, Tsukui T, Abe M, Hosokawa M (2011) The allenic carotenoid fucoxanthin, a novel marine nutraceutical from brown seaweeds. J Sci Food Agric 91:1166–1174
Moore R, Blackman A, Cheuk C, Mynderse J, Matsumoto G, Clardy J, Woodard R, Craig J (1984) Absolute stereochemistries of the aplysiatoxins and oscillatoxin A. J Org Chem 49:2484–2489
Morais T, Inácio A, Coutinho T, Ministro M, Cotas J, Pereira L, Bahcevandziev K (2020) Seaweed potential in the animal feed: a review. J Mar Sci Eng. https://doi.org/10.3390/jmse8080559
Morais T, Cotas J; Pacheco D, Pereira L (2021) Seaweeds compounds: an ecosustainable source of cosmetic ingredients? Cosmetics 8:8. https://doi.org/10.3390/cosmetics8010008
Moroney NC, O'Grady MN, O'Doherty JV, Kerry JP (2012) Addition of seaweed (*Laminaria digitata*) extracts containing laminarin and fucoidan to porcine diets: influence on the quality and shelf-life of fresh pork. Meat Sci 92:423–429. https://doi.org/10.1016/j.meatsci.2012.05.005
Mrčelić et al (2024) Methane reduction in agriculture: using algal feed to reduce methane emissions from livestock. J Agric Sustain 45(2):123–135

Murphy F, Devlin G, Deverell R, McDonnell K (2013) Biofuel production in Ireland-an approach to 2020 targets with a focus on algal biomass. Energies 6:6391–6412

Nagai H, Yasumoto T, Hokama Y (1996) Aplysiatoxin and debromoaplysiatoxin as the causative agents of a red alga *Gracilaria coronopifolia* poisoning in Hawaii. Toxicon 37:753–761

Nagorskaia VP, Reunov AV, Lapshina LA, Ermak IM, Barabanova AO (2008) Influence of kappa/beta-carrageenan from red alga *Tichocarpus crinitus* on development of local infection induced by tobacco mosaic virus in Xanthinc tobacco leaves. Biol Bull 35:310–314

Navarro AE, Portales RF, Sun-Kou MR, Llanos BP (2008) Effect of pH on phenol biosorption by marine seaweeds. J Hazard Mater 156:405–411. https://doi.org/10.1016/j.jhazmat.2007.12.039

Nelson WA (2020) New Zealand seaweeds: an illustrated guide. Te Papa Press, Wellington

Neori A, Chopin T, Troell M, Buschmann AH et al (2004) Integrated aquaculture: rationale, evolution and state of the art emphasizing seaweed biofiltration in modern mariculture. Aquaculture 231:361–391

Neveux N, Bolton JJ, Bruhn A, Roberts DA, Ras M (2018) The bioremediation potential of seaweeds: Recycling nitrogen, phosphorus, and other waste products. In: Barre SL, Bates SS (eds) Blue biotechnology: production and use of marine molecules. Wiley, Hoboken, NJ, USA, pp 217–241

Newton L (1951) Animal and human nutrition from seaweed resources in Europe: uses and potential. Seaweed utilization. Sampson Low, London, pp 31–57

Ngala BM, Valdes Y, dos Santos G, Perry RN, Wesemael WML (2016) Seaweed-based products from Ecklonia maxima and *Ascophyllum nodosum* as control agents for the root-knot nematodes *Meloidogyne chitwoodi* and *Meloidogyne hapla* on tomato plants. J Appl Phycol 28:2073–2082

Nikolaison L, Dahl J, Bech KS, Bruhn A et al (2012) Energy production fom macroalgae. In Proceedings of the 20th European biomass conference. Milan, Italy, pp 18–22

O'Connor I, O'Brien N (1998) Modulation of UVA light induced oxidative stress by beta-carotene, lutein and astaxanthin in cultured fibroblasts. J Dermatol Sci 16:226–230

Ojeda FP, Muñoz AA (1999) Feeding selectivity of the herbivorous fish Scartichthys viridis: effects on macroalgal community structure in a temperate rocky intertidal coastal zone. Mar Ecol Prog Ser 184:219–229

Olanrewaju OS, Tee KF, Nelson Igbineh N (2024) Structural risk and reliability assessment of offshore aquaculture floating system for seaweed cultivation. Mater Sci 6(1). https://doi.org/10.35702/msci.10024

Olasehinde TA, Olaniran AO, Okoh AI (2019) Phenolic composition, antioxidant activity, anticholinesterase potential and modulatory effects of aqueous extracts of some seaweeds on β-amyloid aggregation and disaggregation. Pharm Biol 57(1):460–469. https://doi.org/10.1080/13880209.2019.1634741

Omar HH, Shiekh HM, Gumgumjee NM, El-Kazan MM, El-Gendy AM (2012) Antibacterial activity of extracts of marine algae from the Red Sea of Jeddah, Saudi Arabia. Afr J Biotechnol 11(71):13576–13585

Oualid HA, Abdellaoui Y, Laabd M, El Ouardi M, Brahmi Y, Iazza M, Oualid JA (2020) Eco-efficient green seaweed Codium Decorticatum biosorbent for textile dyes: characterization, mechanism, recyclability, and RSM optimization. ACS Omega 5:22192–22207. https://doi.org/10.1021/acsomega.0c02311

Owusu WA, Marfo SA, Osei H, Supap T, Raphael Idem R (2024) Pelagic *Sargassum* waste as an excellent feedstock for bioethanol production: hydrolysis optimisation and kinetics. Biofuels 1–9

Padmakumar K, Ayyakkannu K (1997) Seasonal variation of antibacterial and antifungal activities of the extracts of marine algae from southern coasts of India. Bot Mar 40:507–515. https://doi.org/10.1515/botm.40.1-6.507

Pagels F, Arias A, Guerreiro A, Guedes AC, Moreira MT (2022) Seaweed cosmetics under the spotlight of sustainability. Phycology 2:374–383. https://doi.org/10.3390/phycology2040021

Pandey A, Pandey S, Pathak J, Ahmed H et al (2017) Mycosporine-like amino acids (MAAs) profile of two marine red macroalgae, *Gelidium* sp. and *Ceramium* sp. Int J Appl Biotechnol Biochem 5:12–21

Park HJ, Lee MK, Park YB, Shin YC, Choi MS (2011) Beneficial effects of Undaria pinnatifida ethanol extract on diet-induced-insulin resistance in C57BL/6J mice. Food Chem Toxicol 49:727–733

Park BG, Shin WS, Oh S, Park GM et al (2017) A novel antihypertension agent, sargachromenol D from marine brown algae, *Sargassum siliquastrum*, exerts dual action as an L-type Ca(2+) channel blocker and endothelin A/B2 receptor antagonist. Bioorg Med Chem 25:4649–4655

Park J, Hojun L, Soyeon C, Lalit P, Stephen D, Jonas D, Joon-Tae P, Taejun H (2021) Extracts of red seaweed, *Pyropia yezoensis*, inhibit melanogenesis but stimulate collagen synthesis. J Appl Phycol 33. https://doi.org/10.1007/s10811-020-02305-y

Park SL (2020) Cosmetic composition for anti-wrinkle activity comprising fermented soybean paste and seaweed extracts as active ingredient. KR20200080500A, 7 July 2020

Paul JP, Muthu S (2016) Screening of analgesic activity of methanolic extract of *Chaetomorpha litorea* harvey (green seaweed) in Koothankuzhi coast. Tamil Nadu, India, pp 59–62

Pavoni B, Caliceti M, Sperni L, Sfriso A (2003) Organic micropollutants (PAHs, PCBs, pesticides) in seaweeds of the lagoon of Venice. Oceanol Acta 26:585–596. https://doi.org/10.1016/S0399-1784(03)00052-5

Peñalver R, Lorenzo JM, Ros G, Amarowicz R, Pateiro M, Nieto G (2020) Seaweeds as a functional ingredient for a healthy diet. Mar Drugs 18(6):301. https://doi.org/10.3390/md18060301. PMID: 32517092; PMCID: PMC7345263

Pena-Rodriguez A, Mawhinney TP, Ricque-Marie D, Cruz-Suarez LE (2011) Chemical composition of cultivated seaweed Ulva clathrata (Roth) C. Agardh. Food Chem 129:491–498

Peng J, Yuan JP, Wu CF et al (2011) Fucoxanthin, a marine carotenoid present in brown seaweeds and diatoms: metabolism and bioactivities relevant to human health. Mar Drugs 9:1806–1828

Pereira L (2018) Biological and therapeutic properties of the seaweed polysaccharides. Int Biol Rev 2(2):200

Pérez MJ, Falqué E, Domínguez H (2016) Antimicrobial action of compounds from marine seaweed. Mar Drugs 14(3):52. https://doi.org/10.3390/md14030052

Pérez-López P, González-García S, Jeffryes C et al (2014) Life cycle assessment of the production of the red antioxidant carotenoid astaxanthin by microalgae: from lab to pilot scale. J Clean Prod 64:332–344

Petrou K, Baker KG, Nielsen DA, Hancock AM, Schulz KG, Davidson AT (2019) Acidification diminishes diatom silica production in the southern ocean. Nat Clim Change 9:781–786. https://doi.org/10.1038/s41558-0190557-y

Pierra M, Trably E, Godon J-J, Bernet N (2014) Fermentative hydrogen production under moderate halophilic conditions. Int J Hydrogen Energy 39:7508–7517

Pimentel FB, Alves RC, Rodrigues F, Beatriz PP (2018) Oliveira macroalgae-derived ingredients for cosmetic industry—an update. Cosmetics 5(2):1–18

Pokoo-Aikins G, Nadim A, El-Halwagi MM, Mahalec V (2010) Design and analysis of biodiesel production from algae grown through carbon sequestration". Clean Technol Environ Policy 12(3):239–254

Polat S, Trif M, Rusu A, Šimat V, Čagalj M, Alak G, Meral R, Özogul Y, Polat A, Özogul F (2021) Recent advances in industrial applications of seaweeds. Crit Rev Food Sci Nutr. https://doi.org/10.1080/10408398.2021.2010646

Polat S, Ozogul Y (2013) Seasonal proximate and fatty acid variations of some seaweeds from the northeastern Mediterranean coast. Oceanologia 55(2):375–391

Ponnanikajamideen M, Suneetha V, Rajeshkumar S (2017) Antidiabetic, antihyperlipedimic and antioxidant activity of marine brown seaweed Padina tetrastromatica. J Chem Pharm Sci 10(1):379–384

Pourakbar L, Moghaddam SS, El Enshasy HA, Sayyed RZ (2021) Antifungal activity of the extract of a macroalgae, Gracilariopsis persica, against four plant pathogenic fungi. Plants 10(9):1781. https://doi.org/10.3390/plants10091781

Prakash S, Ravikumar S, Reddy KVR, Kannapiran E (2014) Spermicidal activity of Indian seaweeds: an in vitro study. Andrologia 46:408–416

Prisa D, Fresco R, Jamal A, Saeed MF, Spagnuolo D (2024) Exploring the potential of macroalgae for sustainable crop production in agriculture. Life 14:1263. https://doi.org/10.3390/life14101263

Probst L, Frideres L, Pedersen B, Amato F (2015) Sustainable, safe and nutritious food: new nutrient sources. Publications Office of the European Union, Luxembourg, p 17

Punampalam R, Khoo KS, Sit NM (2018) Evaluation of antioxidant properties of phycobiliproteins and phenolic compounds extracted from Bangia atropurpurea. Malay J Fundam Appl Sci 14(2):289–297

Quitral V, Sepúlveda M, Gamero-Vega G, Jiménez P (2022) Seaweeds in bakery and farinaceous foods: a mini-review. Int J Gastron Food Sci 28:100403. https://doi.org/10.1016/j.ijgfs.2021.100403

Radmer RJ (1996) Algal diversity and commercial algal products. Bioscience 46:263–270

Rahbari H, Akram A, Pazoki M, Aghbashlo M (2019) Bio-oil production from *Sargassum* macroalgae: a green and healthy source of energy. Jundishapur J Health Sci 11(1):e84301. https://doi.org/10.5812/jjhs.84301

Rakib MRJ, Jolly YN, Dioses-Salinas DC, Pizarro-Ortega CI, De-la-Torre GE, Khandaker MU, Alsubaie A, Almalki ASA, Bradley DA. (2021). Macroalgae in biomonitoring of metal pollution in the Bay of Bengal coastal waters of Cox's Bazar and surrounding areas. Sci Rep 11(1):20999. https://doi.org/10.1038/s41598-021-99750-7

Ramamoorthi L, Jeyabalan S, Sankar S, Begum MY, Duraipandian C, Sekar M, Wong LS, Subramaniyan V (2025) Anti-arthritic potential of crude sulfated polysaccharide from marine macroalgae *Sargassum ilicifolium* (Turner) C. Agardh: Regulation of cytokine cascade. Biomol Concepts 16(1). https://doi.org/10.1515/bmc-2022-0050. PMID: 39846213

Rani S, Kumar AK (2021) Iodiesel production from macroalgae *Cladophora glomerata*. Int Adv Res J Sci Eng Technol 8(11):56–59

Ravikumar S, Inbaneson S, Suganthi P, Gokulakrishnan R, Venkatesan M (2011) In vitro antiplasmodial activity of ethanolic extracts of seaweed macroalgae against *Plasmodium falciparum*. Parasitol Res 108:1411–1416

Rengasamy KR, Amoo SO, Aremu AO et al (2015) Phenolic profiles, antioxidant capacity, and acetylcholinesterase inhibitory activity of eight South African seaweeds. Appl Phycol 27(4):1599–1605. https://doi.org/10.1007/s10811-014-0438-8

Renita A, Davuluri D (2015) Extraction of dye from marine macroalgae. Int J ChemTech Res 8(3):1060–1063

Rhein-Knudsen N, Ale MT, Meyer AS (2015) Seaweed hydrocolloid production: an update on enzyme assisted extraction and modification technologies. Mar Drugs 13:3340–3359

Ross F, Tarbuck P, Macreadie PI (2022) Seaweed afforestation at largescales exclusively for carbon sequestration: critical assessment of risks, viability and the state of knowledge. Front Mar Sci 9:1015612. https://doi.org/10.3389/fmars.2022.1015612

Russell LM, Rasch PJ, Mace GM, Jackson RB, Shepherd J, Liss P et al (2012) Ecosystem impacts of geoengineering: a review for developing a science plan. Ambio 41:350–369. https://doi.org/10.1007/s13280-012-0258-5

Saifullah M, Stephen MM, Khattak B (2007) Isolation of Trichoderma harzianum and in vitro screening for its effectiveness against root-knot nematodes (*Meloidogyne* sp.) from Swat, Pakistan. Pak J Nematol 25:313–322

Sakakibara H (2006) Cytokinins: activity, biosynthesis, and translocation. Annu Rev Plant Biol 57:431–449

Salama AJ, Sathianeson Satheesh S, Balqadi AA (2017) Antifouling activities of methanolic extracts of three macroalgal species from the Red Sea. J Appl Phycol 17:1345–1346

Saldarriaga-Hernandez S, Hernandez-Vargas G, Iqbal MN et al (2020) Bioremediation potential of Sargassum sp. biomass to tackle pollution in coastal ecosystems: circular economy approach. Sci Total Environ 715:136978

Salem DMSA, Ismail MM, Aly-Eldeen MA (2019) Biogenic synthesis and antimicrobial potency of iron oxide (Fe_3O_4) nanoparticles using algae harvested from Mediterranean Sea, Egypt. Egypt J Aquat Res 45:197–204. https://doi.org/10.1016/j.ejar.2019.07.002

Salem DMS, Ismail MM, Tadros HRZ (2020) Evaluation of the antibiofilm activity of three seaweed species and their biosynthesized iron oxide nanoparticles (Fe_3O_4-NPs). Egypt J Aquat Res 46(4):333–339. https://doi.org/10.1016/j.ejar.2020.09.001

Salem DMSA, Ismail MM (2021) Characterization of cellulose and cellulose nanofibers isolated from various seaweed species. Egypt J Aquat Res 48(4):307–313. https://doi.org/10.1016/j.ejar.2021.11.001

Sathasivam R, Radhakrishnan R, Hashem A, Abd Allah EF (2017) Microalgae metabolites: a rich source for food and medicine. Saudi J Biol Sci 3:1–14

Satheesh S, Baakdah MA, Al-Sofyani AA (2016) Natural antifouling compound production by microbes associated with marine macroorganisms. a review. Electron J Biotechnol 21:26–35

Schenk PM, Thomas-Hall SR, Stephens E, Marx UC et al (2008) Second generation biofuels: high efficiency microalgae for biodiesel production. Bioenergy Res 1:20–43

Scheuer PJ, Haman MT, Gravalos DG (2000) Cytotoxic and antiviral compound. U.S. Patent 6011010

Ścieszka S, Klewicka E (2019) Algae in food: a general review. Crit Rev Food Sci Nutr 59:3538–3547. https://doi.org/10.1080/10408398.2018.1496319

Seo Y-B, Lee Y-W, Lee C-H, You H-C (2010) Red algae and their use in papermaking. Biores Technol 101:2549–2553

Seung-Hong L, You-Jin J (2013) Anti-diabetic effects of brown algae derived phlorotannins, marine polyphenols through diverse mechanisms. Fitoterapia 86:129–136

Shabarish S, Tamilarasan K, Rajesh Banu J et al (2023) Anaerobic fermentation of seaweed for enhanced biohydrogen production through combined sonic surfactant disintegration: process optimization and energy assessment. Biomass Conv Bioref. https://doi.org/10.1007/s13399-023-03898-3

Shama A, Joyce SG, Mari FD, Ranga Rao A, Ravishankar GA, Hudaa N (2019) Macroalgae and microalgae: novel sources of functional food and feed. In: Ravishankar GA, Ranga Rao A (eds) Handbook of algal technologies and phytochemicals. Food, health and nutraceutical applications, vol I. CRC Press, USA, pp 207–219

Shanmuganathan B, Malar DS, Sathya S, Devi KP (2015) Antiaggregation potential of *Padina gymnospora* against the toxic Alzheimer's beta-amyloid peptide 25–35 and cholinesterase inhibitory property of its bioactive compounds. PLoS ONE 10:e0141708

Sharma B, Purkayastha DD, Hazra S et al (2015) Biosynthesis of gold nanoparticles using a fresh water green alga, *Prasiola crispa*. Mater Lett 116:94–97

Sharma A, Sharma S, Sharma K, Chetri SPK, Vashishtha A, Singh P, Kumar R, Rathi B, Agrawal V (2016) Algae as crucial organisms in advancing nanotechnology: a systematic review. J Appl Phycol 28:1759–1774. https://doi.org/10.1007/s10811-015-0715-1

Sharma PP, Baskaran V (2021) Polysaccharide (laminaran and fucoidan), fucoxanthin and lipids as functional components from brown algae (*Padina tetrastromatica*) modulates adipogenesis and thermogenesis in diet-induced obesity in C57BL6 mice. Algal Res 54:102187. https://doi.org/10.1016/j.algal.2021.102187

Shi H, Zheng W, Zhang X, Zhu M, Ding D (2015) Ecological-economic assessment of monoculture and integrated multi-trophic aquaculture in Sanggou Bay of China. Aquaculture 410:172–178

Shi Q, Wang A, Lu Z, Qin C et al (2017) Overview on the antiviral activities and mechanisms of marine polysaccharides from seaweeds. Carbohyd Res 453–454:1–9

Shobier AH, Ismail MM, Hassan SW (2023) Variation in anti-inflammatory, anti-arthritic, and antimicrobial activities of different extracts of common Egyptian seaweeds with an emphasis on their phytochemical and heavy metal contents. Biol Trace Element Res 201(4):2071–2087. https://doi.org/10.1007/s12011-022-03297-1

Singh S, Kumar MS (2022) Antiviral applications of macroalgae. In: Ranga Rao A, Ravishankar GA (eds) Sustainable global resources of seaweeds volume 2. Springer, Cham. https://doi.org/10.1007/978-3-030-92174-3_25

Singh J, Gu S (2010) Commercialization potential of microalgae for biofuels production. Renew Sustain Energy Rev 14:2596–2610

Sivagurunathan P, Kumar G, Kobayashi T, Xu K, Kim S (2017) Effects of various dilute acid pretreatments on the biochemical hydrogen production potential of marine macroalgal biomass. Int J Hydrogen Energy 11(45):27600–32760

Smyrniotopoulos V, Merten C, Kaiser M, Tasdemir D (2017) Bifurcatriol, a new antiprotozoal acyclic diterpene from the brown alga *Bifurcaria bifurcata*. Mar Drugs 15:245

Soliman ASh, Ahmed AY, Abdel-Ghafour SE, El-Sheekh MM, Sobhy HM (2018) Antifungal bio-efficacy of the red algae *Gracilaria confervoides* extracts against three pathogenic fungi of cucumber plant. Middle East J Appl Sci 08(03):727–735

Stiger-Pouvreau V, Bou.rgougnon N, Deslandes E (2016) Carbohydrates from seaweeds. In: Fleurence J, Levine I (eds) Seaweed in health and disease prevention. Elsevier, San Diego

Suganya T, Nagendra Gandhi N, Renganathan S (2013) Production of algal biodiesel from marine seaweeds *Enteromorpha compressa* by two step process: optimization and kinetic study. Bioresour Technol 128:392–400

Sukrri NNA, Farizan AF, Ramzi MM, Rawi NN, Abd Rahman NI, Bakar K, Siong JYF, Azemi AK, Ismail N (2024) Antifouling activity of Malaysian green seaweed Ulva lactuca and its isolated non-polar compound. Heliyon 10(19):38366. https://doi.org/10.1016/j.heliyon.2024.e38366

Suresh J, Subbaiyan R, Ganesan A, Ramasubramanian B (2024) Effective antifouling action of seaweed metabolite waste against marine foulers. Sustain Chem One World 2. https://www.sciencedirect.com/science/article/pii/S2950357424000040

Thennarasan S, Murugesan S, Chidambaranathan N, Sivamurugan V (2016) Analgesic, anti-inflammatory and antipyretic activity of the methanol extracts of brown alga *Lobophora variegate* (J.V. Lamouroux) Womersley ex E.C. Oliveir. Am J Phytomed Clin Therap 4(02):042–057

Thiyagarasaiyar K, Goh B-H, Jeon Y-J, Yow Y-Y (2020) Algae metabolites in cosmeceutical: an overview of current applications and challenges. Mar Drugs 18:323. https://doi.org/10.3390/md18060323

Thompson TM, Young BR, Baroutian S (2019) Advances in the pretreatment of brown macroalgae for biogas production. Fuel Process Technol 195:106151. https://doi.org/10.1016/j.fuproc.2019.106151

Tolasa S, Cakli S, Ostermeyer U (2005) Determination of astaxanthin and canthaxanthin in salmonid. Eur Food Res Technol 221:787–791

Torres MD, Kraan S, Dominguez H (2020) Sustainable seaweed technologies cultivation, biorefinery, and applications. https://www.sciencedirect.com/book/9780128179437/sustainable-seaweedtechnologies?via=ihub=

Truus K, Vaher M, Taure I (2001) Algal biomass from *Fucus vesiculosus* (Phaeophyta): investigation of the mineral and alginate components. Proc Est Acad Sci 50:95–103

Ulagesan S, Nam TJ, Choi YH (2021) Extraction and purification of R-phycoerythrin alpha subunit from the marine Red Algae *Pyropia yezoensis* and its biological activities. Molecules 26(21):6479. https://doi.org/10.3390/molecules26216479. PMID: 34770894; PMCID: PMC8587297

Urbanová T, Tarkowská D, Novák O, Strnad M, Hedden P (2013) Analysis of gibberellins as free acids by ultra performance liquid chromatography-tandem mass spectrometry. Talanta 112:85–94

Usoltseva RV, Shevchenko NM, Malyarenko OS, Ishina IA et al (2018) Structure and anticancer activity of native and modified polysaccharides from brown alga *Dictyota dichotoma*. Carbohyd Polym 180:21–28

van den Burg SW, Röckmann C, Banach JL, van Hoof L (2020) Governing risks of multi-use: seaweed aquaculture at offshore wind farms. Front Mar Sci (International Institute of Fisheries Economics and Trade) 7:60. https://doi.org/10.3389/fmars.2020.00060
Van De Velde F, Peppelman HA, Rollema HS, Hans R (2001) On the structure of κ/ι-hybrid carrageenans. Carbohyd Res 331:271–283
Van den Burg SW, van Duijn AP, Bartelings H, vanKrimpen MM, Poelman M (2016) The economic feasibility of seaweed production in the North Sea. Aquacult Econ Manage 20:235–252. https://doi.org/10.1080/13657305.2016.1177859
van der Wal H, Sperber BL, Houweling-Tan B, Bakker RR et al (2013) Production of acetone, butanol and ethanol from biomass of the green seaweed *Ulva lactuca*. Bioresour Technol 128:431–437
Vaughan VC, Hassing MR, Lewandowski PA (2013) Marine polyunsaturated fatty acids and cancer therapy. Br J Cancer 108(3):486–492
Vega-Moreno D, Sosa-Ferrera Z, Santana-Rodríguez JJ (2007) SPME and SPE comparative study for coupling with microwave-assisted micellar extraction in the analysis of organochlorine pesticides residues in seaweed samples. Microchem J 87:139–146. https://doi.org/10.1016/j.microc.2007.07.002
Venkateswarlu U (2015) Ultrasonic studies on sargassum dye using water/ethanol extraction at different temperatures. Int Res J Eng Technol 2(8):1130–1138
Venkatraman A, Yahoob SAM, Nagarajan Y, Harikrishnan S et al (2018) Pharmacological activity of biosynthesized gold nanoparticles from brown algae- seaweed *Turbinaria conoide*. NanoWorld J 4(1):17–22
Villas-Bôas AB, Figueiredo MAO (2004) Are antifouling effects in coralline algae species specific. Braz J Oceanogr 52:11–18
Vinutha B, Prashanth D, Salma K, Sreeja SL,. Pratiti D, Padmaja S, Radhika R, Amit A, Venkateshwarlu K, Deepak M (2007) Screening of selected Indian medicinal plants for acetylcholinesterase inhibitory activity. J Ethnopharmacol 109:359–363
Visch W, Kononets M, Hall PO, Nylund GM, Pavia H (2020) Environmental impact of kelp (Saccharina latissima) aquaculture. Mar pollut Bull 155:110962. https://doi.org/10.1016/j.marpolbul.2020.110962
Wang SB, Jia YH, Wang LH, Zhu FH, Lin YT (2013a) Enteromorpha prolifera supplemental level: effects on laying performance, egg quality, immune function and microflora in feces of laying hens. Chin J Anim Nutr 25:1346–1352
Wang J, Jin W, Hou Y, Niu X, Zhang H, Zhang Q (2013b) Chemical composition and moisture-absorption/retention ability of polysaccharides extracted from five algae. Int J Biol Macromol 57:26–29
Wang SB, Shi XP, Zhou CF, Lin YT (2013c) Enteromorpha prolifera: effects on performance, carcass quality and small intestinal digestive enzyme activities of broilers. Chin J Anim Nutr 25:1332–1337
Wang W-L, Fernández-Méndez M, Elmer F, Gao G, Zhao Y, Han Y, Li J, Chai F, Dai M (2023) Ocean afforestation is a potentially effectiveway to remove carbon dioxide. Nat Commun. https://doi.org/10.1038/s41467-021-22837-2
WHO (1989) Toxicological evaluation of certain food additives and contaminants WHO Food Additives Series 24. World Health Organization, Geneva. http://www.inchem.org/documents/jecfa/jecmono/v024je11.htm
Wilcox HA (1975) Hothouse Earth. Praeger, New York
Wu L, Sun J, Su X, Yu Q et al (2016) A review about the devel-opment of fucoidan in antitumor activity: progress and challenges. Carbohyd Polym 154:96–111
Wyrtki K (1962) The oxygen minima in relation to ocean circulation. In: Deep Sea research and oceanographic abstracts. Elsevier, pp 11–23. https://doi.org/10.1016/00117471(62)90243-7
Xiao X, Agusti S, Lin F, Li K, Pan Y, Yu Y, Zheng Y, Wu J, Duarte CM (2017) Nutrient removal from Chinese coastal waters by large-scale seaweed aquaculture. Sci Rep 7:46613. https://doi.org/10.1038/srep46613

Xiao X, Agustí S, Yu Y, Huang Y, Chen W, Hu J et al (2021) Seaweed farms provide refugia from ocean acidification. Sci Total Environ 776:145192. https://doi.org/10.1016/j.scitotenv.2021.145192

Yadav M, Thakore S, Rajendrasinh Jadeja R (2022) Removal of organic dyes using *Fucus vesiculosus* seaweed bioadsorbent an ecofriendly approach: equilibrium, kinetics and thermodynamic studies. Environ Chem Ecotoxicol 4:67–77. https://doi.org/10.1016/j.enceco.2021.12.003

Yanagisawa M, Kawai S, Murata K (2013) Strategies for the production of high concentrations of bioethanol from seaweeds production of high concentrations of bioethanol from seaweeds. Bioengineered 4(4):224–235

Yang X-D, Liu C-G, Tian Y-J, Gao D-G, Li W-S, Ma H-L (2017) Inhibitory effect of fucoidan on hypoglycemia in diabetes mellitus anim. Int J Clin Exp Med 10(5):8529–8534

Yong WTL, Thien VW, Rennielyn Rupert R, Rodrigues KF (2022) Seaweed: a potential climate change solution. Renew Sustain Energy Rev 159:112222. https://doi.org/10.1016/j.rser.2022.112222

Young MO (2015) Marine animal entanglements in mussel aquaculture gear (Ísafjörður, Iceland: University of Akureyri)

Yusaf T, Goh S, Borserio J (2011) Potential of renewable energy alternatives in Australia. Renew Sustain Energy Rev 1:2214–2221

Zárate R, Portillo E, Teixidó S, de Carvalho MAAP, Nunes N, Ferraz S, Seca AML, Rosa GP, Barreto MC (2020) Pharmacological and cosmeceutical potential of seaweed beach-casts of Macaronesia. Appl Sci 10:5831

Zava TT, Zava DT (2011) Assessment of Japanese iodine intake based on seaweed consumption in Japan: a literature-based analysis. Thyroid Res 4:14

Zhang S, Zhang S, Wang H, Chen Y (2006) Characteristics of sweetpotato stem nematode in China. Acta Phytopathol Sin 36:22–27

Zhao Y, Bourgougnon N, Lanoisellé J-L, Lendormi T (2022) Biofuel production from seaweeds: a comprehensive review. Energies 15:9395. https://doi.org/10.3390/en15249395

Zhou M, Hünerberg M, Chen Y, Reuter T, McAllister TA, Evans F, Critchley AT, Guan LL (2018) Air-dried brown seaweed, Ascophyllum nodosum, alters the rumen microbiome in a manner that changes rumen fermentation profiles and lowers the prevalence of foodborne pathogens. Sphere 3(1). https://doi.org/10.1128/mSphere.00017-18

Zimmermann MB (2012) The effects of iodine deficiency in pregnancy and infancy. Paediatr Perinat Epidemiol 26(1):108–117

Chapter 4
Different Technologies Used for Isolation of Bioactive Compounds from Seaweed

4.1 Introduction

The extraction of seaweed is the first and most important step in purifying compounds from reaction mixtures or natural sources, either in batch or continuously (Budhiraja 2004). The most crucial stage in obtaining bioactive chemicals is extraction, which involves using a specific amount of solvent to separate compounds from solids or liquids. In order for this method to dissolve and extract the target compound, isolating the solute of interest from the raw material, a solvent that can permeate the solid matrix is needed (Zhang et al. 2018). This process removes the desired substance without removing other elements. The extraction techniques, operating parameters, and pretreatment methods are crucial in determining the quality and yield of the extract (Daud et al. 2022). The choice of method and extraction efficiency is influenced by the seaweed species being analyzed, the type of solvent, temperature, time, pH, environmental impact, cost, quantity, and desired properties (Jacobsen et al. 2019; Lim et al. 2021). The quantity and quality of bioactive chemicals are significantly influenced by the extraction process. Extraction procedures not only affect purity and efficacy, but also production costs and biofunctional activity. Selecting the proper extraction process with optimal operating parameters is crucial for extracting bioactive chemicals from seaweeds, enhancing their bioavailability and potential, and maximizing their therapeutic effects (Park et al. 2024). Various extraction methods have been used to extract these target chemicals, which are classified into two types: conventional extraction techniques and novel extraction techniques (Table 4.1). Several aspects must be considered when selecting an extraction process, including time, recovery rate, cost, feasibility of scaling up, volume, and type of solvents (Perez-Vazquez et al. 2023).

G. M. El Zokm et al., *Rethinking Seaweeds via Applications and Technologies*,
Environmental Science and Engineering, https://doi.org/10.1007/978-3-032-14956-5_4

Table 4.1 Comparison between conventional and green techniques (Own elaboration)

	Conventional techniques	Green techniques
Types	Solid-liquid extraction (SLE)	• Ultrasound-assisted extraction (UAE) • Enzyme-assisted extraction (EAE) • Pressured liquid extraction (PLE) • Field-based technology and supercritical fluid extraction (SFE)
Aims	Using organic solvents with a certain extraction time	Using a small amount of solvent under specific conditions with a fast extraction time
Advantages	Simple	• Does not use organic solvents but has high yield • Short extraction time with a relatively small amount of solvent
Disadvantage	Long extraction time, using a lot of solvents, acids and bases resulted in the degradation of polysaccharide	Extraction results depend on the time, pH, enzyme and temperature

4.2 Conventional Extraction Methods

Conventional extraction methods are widely commonly used worldwide to extract biocompounds from seaweed, but they are not environmentally friendly. These approaches, which include mechanical grinding, centrifugation, chemical procedures and enzyme-based methods, are commonly used to extract bioactive chemicals from seaweeds, despite their limitations. Physical methods are slow and ineffective in terms of output. Chemical methods using organic solvents like ethanol, methanol, n-hexane and acetone can have negative impacts on the active constituents, decrease extraction yield, raise assembly costs, and pose protection risks due to residual solvent (Park et al. 2024). Successful extraction using conventional techniques often requires a heating process that depends on convection of the solvent and subsequent conduction of the matrix particles (Grosso et al. 2015). Steam distillation, infusion, percolation, maceration, and Soxhlet extraction are a few examples of traditional techniques (Bordoloi and Goosen 2020). The extraction process presents many difficulties at the industrial level, especially with large quantities. In order to overcome these constraints, effective and ecologically friendly extraction methods must be developed. Traditional techniques usually use organic, alkaline, or acidic solvents for solid-liquid extraction (SLE) at room temperature or under higher atmospheric pressure (Quitério et al. 2022). These techniques frequently require many hours to extract the bioactive components from seaweed. The extraction yield and quality depend on the type of bioactive compounds and the seaweed species. These methods typically involve the use of organic solvents, which require large volumes and extended extraction times (Khalil et al. 2018). For phycocolloids, hot water is the common solvent due to their hydrocolloids' water solubility (Zollmann et al. 2019). For example, alginate and alginic acid are extracted through hot alkali extraction, which is an aqueous-based and converts insoluble alginate salts into water-soluble alginate salts. Other solvents, such as acids or organic solvents are frequently employed, depending on the specific chemical being targeted. Conventional extractions typically involve a heating method that combines convection through the solvent and conduction from the surfaces to the core of the matrix particles (Grosso et al. 2015). Additionally, optimizing the extraction process is challenging due to variables including raw material size, temperature, pH, seaweed species, collection season, geographic site, and concentration of solvent (Khalil et al. 2018). One cost-effective extraction method that requires minimal equipment is maceration. The solubility of the target compound in this method is determined by temperature and agitation. The procedures employed in this method can also be easily modified to yield a variety of chemicals of interest. The ability to utilized different solvents, temperatures, and agitation levels enhances adaptability and can enhance the efficiency and selectivity of mass transfer. However, the chemicals must be separated from the matrix through multiple cycles of centrifugation or filtration (Garcia-Vaquero et al. 2020). The extraction method known as percolation works continuously compared to maceration. It results in higher efficiency because a fresh solvent continuously replaces the saturated one. Although heating can be used, the usual operating conditions are atmospheric pressure at room temperature. The efficiency of reflux extraction is higher than that of maceration and percolation, despite slightly lower extraction time and solvent requirements. The operational parameters for reflux extraction are atmospheric pressure and heating (Zhang et al. 2018). The extraction of phytochemicals may not always be effectively accomplished by traditional extraction methods like solvent extraction. This is due to the fact that some compounds are inaccessible to solvents in a standard extraction method, either because they are disseminated in the cytoplasm of the cell or are kept in the cell wall through hydrogen or hydrophobic interaction (Fleurence et al. 1995). On the other hand, conventional Solid Liquid Extraction (SLE) has drawbacks such as post-concentration/separation, prolonged extraction times, substantial solvent use, and selectivity issues (Quitério et al. 2022). In conventional extraction methods solvents, especially organic solvents have historically caused environmental issues. This is due to their classification as Volatile Organic Compounds (VOCs), which are organic pollutants that contribute to the formation of photochemical smog in the troposphere and ozone depletion in the stratosphere by acting as a source of radicals (Calvo-Flores 2020).

Liquid-Liquid Extraction (LLE) requires large volumes of organic solvents, leading to the production of hazardous waste. Therefore, using minimal amounts of organic solvents and biodegradable materials is a more environmentally friendly extraction option (Sanagi et al. 2016). There are several drawbacks to solvent extraction, including the need for large amounts of organic solvents, extended extraction times, and issues with selectivity (Dulanlebit and Hernani 2023).

Nowadays, the main priorities for the industry are quality, cost-effectiveness, and environmental sustainability. The goal of green or alternative extraction methods is to increase extraction yields while reducing or eliminating the usage of solvents, energy consumption, and extraction time, Green or alternative extraction techniques aim to boost extraction yields while lowering or doing away with the need for solvents, energy, and extraction time, which is typically only a few minutes. The development of sustainable extraction methods is essential to raising the seaweed industry's potential for value addition. The advancement of sustainable extraction techniques is crucial for enhancing the value-added potential of the rapidly growing seaweed industry (Bordoloi and Goosen 2020).

4.3 Alternative Extraction Methods (Green Technologies)

As previously mentioned, traditional extraction methods employ severe processing conditions that can produce significant volumes of hazardous waste, cause environmental issues, and leave a significant carbon imprint. Furthermore, in order to apply safe end products and adhere to food and pharmaceutical sector laws, downstream processing is necessary if the solvent being utilized is toxic (Bordoloi and Goosen 2020). Using environmentally friendly solvents, including ionic liquids, eutectic solvents, surfactants, or solvents made from biomass, is another way to reduce the adverse effects of the extraction procedure (Matos et al. 2021). Technologies that are nonchemical, quick, energy-efficient, zero-waste, non-hazardous, environmentally benign, inexpensive, adaptable, suitable for untreated biomass, and able to be integrated with other green technologies are referred to as smart technologies (Zollmann et al. 2019). Green extraction techniques remain popular in the food sector (Chemat et al. 2017). The innovation and design of "green method" extraction using organic solvents, selectivity, lower temperature, and short extraction time are advantages of these methods (Vijayan et al. 2014). Green extraction is a more beneficial method for extracting seaweed hydrocolloids without the use of chemicals (Yew et al. 2016). There are several new and alternative extraction techniques that are environmentally friendly. These techniques include field-based technologies, including microwave-assisted extraction (MAE) and radiofrequency-assisted extraction, pressurized liquid extraction (PLE), ultrasound-assisted extraction (UAE), enzyme-assisted extraction (EAE), and supercritical fluid extraction (SFE) (Muhammad et al. 2023). The combination of these methods has resulted in better seaweed extract. Environmentally friendly, economical, fast, sensitive, and accurate extraction methods are preferred over conventional extraction (Lim et al. 2021). As shown in Table 4.2, the main alternative methods for removing various biocompounds from seaweeds and the results that they produce.

4.3.1 Ultrasound-Assisted Extraction (UAE)

Ultrasound-assisted extraction (UAE) involves the use of sound waves with frequencies greater than 20 kHz to create pressure differences that form bubbles through ultrasonic vibrations ranging from 20 to 100 kHz. When these bubbles collapse and cavitate, they break the liquid-solid barrier near the particles, releasing bioactive

Table 4.2 Various alternative extraction methods used for extracted bioactive compounds from seaweeds (Own elaboration)

Seaweed specie	Compounds	Extraction process	Extraction conditions	Yield	References
Laminaria japonica	Carotenoids	UAE	30 min; MeOH; 1:20	0.336 g/kg	Lu et al. (2013)
	Chlorophyll a			3.028 g/kg	
Ascophyllum nodosum	Total phenolic		25 min; 0.03 M HCl; 114 μM	143.12 mgGAE/g	Kadam et al. (2015)
	Fucose			87.06 mg/g	
	Uronic acids			128.54 mg/g	
Ecklonia cava	Fucoidans	EAE	Celluclast 24 h; 50 °C	1.8%	Lee et al. (2012)
Fucus vesiculosus	Fatty acids	SFE	304 bar; 60 °C; 60 min	21.9 mg/g	Bogolitsyn et al. (2017)
	Polyphenols			20.2 mg/g	
Gelidium sesquipedale	Protein	SWE	185 °C; 2 mL/min; 240 min	150 mg/g	Trigueros et al. (2021)
Ulva lactuca	Protein	PEF	Electric field strength of 7.5 kV/cm with 0.05 ms pulses	15%	Postma et al. (2018)
	Carbohydrates			< 15%	
Gracilaria vermiculophylla	Agar	MAE	10–20 min; 70–90 °C	14.4%	Sousa et al. (2021)
Eisenia bicyclis	Carotenoid (Fucoxanthin)	PSE	103.4 bar; 110 C; 5 min; 2 g; EtOH 90%; static	0.42 mg/g	Shang et al. (2011)

substances into the matrix. The effectiveness of ultrasound in causing cavitation is influenced by factors such as ultrasonic frequency and intensity, the viscosity and surface tension of the medium, as well as surrounding environmental factors like pressure and temperature (Tiwari 2015). In solid-liquid suspensions, asymmetrical bubbles are formed for duping vapor from the solvent. This leads to subsidence and the generation of mechanical power through micro turbulence, ultimately breaking the cell wall of algae and improving extraction efficacy (Zollmann et al. 2019; Bordoloi and Goosen 2020). To optimize this extraction method, parameters such as frequency, power, temperature, time, and sample to solvent ratio need to be carefully adjusted. The use of ultrasound in extraction processes has been shown to reduce processing costs, solvent usage, and extraction time (Ummat et al. 2021).

UAE technique reduces extraction times and increases extraction yield. Compared to alternative techniques, it is an economical, easy, eco-friendly, and efficient extraction technology. Because of its cheap maintenance costs, small equipment size, ease of industrial scale-up, and fewer process stages, it has been widely adopted in the food sector (Khalil et al. 2018; Bordoloi and Goosen 2020). This technique has proven successful in extracting pigments, polysaccharides, and phenolic compounds from various macroalgae species (Carreira-Casais et al. 2021). In this context, Suchintita Das et al. (2021) reported that UAE technology has emerged as a clean, green, and efficient method of extracting protein compared to conventional methods. The UAE method is more efficient than traditional extraction because it yields a higher total phenolic compound and antioxidant activity using low temperature, short time, and a small amount of solvent (Dang et al. 2017a). The UAE method produces higher chlorophyll and carotenoid content compared to the conventional method from *Ulva flexuosa* (Fabrowska et al. 2018). Additionally, fucoxanthin pigment from the brown alga *Padina* species is characterized by high antioxidant activity using the UAE method (Dang et al. 2017b). In comparison to traditional extraction methods, Rafiquzzaman et al. (2016) demonstrated that combining UAE with aqueous and alkaline solvents increased the yield and purity of carrageenan (Fig. 4.1). Additionally, Kadam et al. (2015) found that, compared to traditional approaches, laminarin extraction using UAE from the brown alga *Ascophyllum nodosum* resulted in a higher amount and antioxidant ability. However, a main disadvantage is that UAE requires a substantial initial capital investment for industrial use because of the expense of the equipment and energy intake, and its application is currently restricted. Nevertheless, it can be an interesting technology when used alone or in combination with other extraction methods (Bordoloi and Goosen 2020).

4.3.2 Enzyme-Assisted Extraction (EAE)

In this technique, commercial digestive enzymes such as proteases and carbohydrases were used to break down seaweed tissues and help release a variety of bioactive compounds from the seaweeds (Hardouin et al. 2014a, b). The bioactive components were released more effectively through enzymatic extraction compared to traditional water extraction. Additionally, enzymes can convert water-insoluble materials in the seaweeds into water-soluble materials, offering advantages over the use of organic solvents. The overall yield showed that enzymatic extraction recovered about 1.2 to 3 times more cell wall material compared to control water extraction in different seaweeds. The enzymes break down the algal cell walls and complex inner storage materials, thereby releasing compounds into solution which may explain the higher yield of enzymatic extracts over the conventional water extracts. Enzymatic extraction is able to maintain the structure and function of seaweed polysaccharides (Rhein-Knudsen et al. 2015). In addition to using microorganisms, enzymatic extraction can be done by combining carrageenan with cellulose to extract the protein in seaweed (Rhein-Knudsen et al. 2015). To increase specificity and selectivity in this procedure, the ideal enzyme reaction conditions and enzyme concentration should be adjusted (Mena-García et al. 2019). Since it uses no hazardous chemicals or organic solvents, this sustainable extraction technique can be regarded as the greenest of all, making

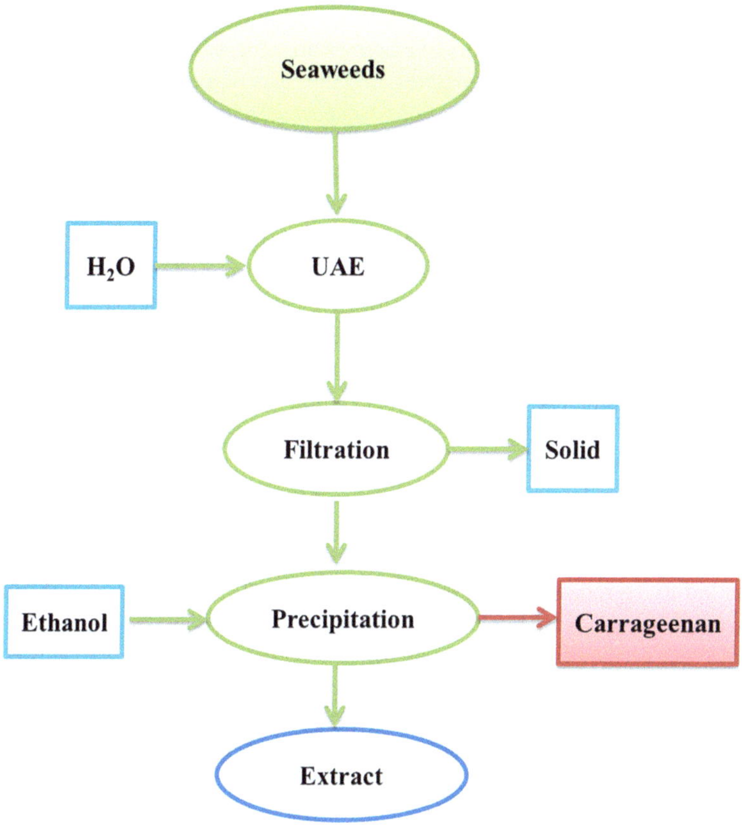

Fig. 4.1 Ultrasound-assisted method for the extraction of carrageenan (Own elaboration)

it a very selective and specific method. Additionally, it is low cost, converting water-insoluble materials into water-soluble materials which has advantages over the use of organic solvents.

The bioactivity varies according to the seaweed species and the extraction enzymes used. To obtain unique bioactivity, it is essential to choose and optimize the extraction conditions for each species of seaweed (Habeebullah et al. 2020). The efficiency of targeted biocompounds extraction can be improved by several critical parameters including pH, temperature, the solvent used, substrate-to-solvent ratio, and mixing conditions (Muhammad et al. 2023). In order to improve the extraction efficiency of bioactive components, seaweed biorefineries have extensively utilized enzymes as a potential biotechnological application (Vásquez et al. 2019). In this case, several seaweeds have had their phlorotannins, proteins, or hydrocolloids extracted using the EAE method (Habeebullah et al. 2020). Additionally, it is employed to extract carrageenan from red seaweeds, with strong gelling qualities and a high extraction yield (Khalil et al. 2018). Further research has demonstrated that EAE preserves the structural integrity of the target substances, including proteins, lipids, and various polysaccharides like ulvans and fucoidans (Hardouin et al. 2016; Bordoloi and Goosen 2020). Only a few numbers of enzymes are accessible for industrial application, particularly for seaweed substrates, and the ones that are still quite expensive. Thus, the majority of EAE applications to seaweeds have made use of enzymes from terrestrial biomass. High glucose yields, for instance, could be obtained by using cellulases to release glucose (fermentable sugars). However, because terrestrial biomass enzymes have sluggish enzymatic kinetics and low substrate specificity, the EAE process might be boosted and extraction efficiency increased by using the potential of new enzymes from marine sources (Bordoloi and Goosen 2020).

The EAE method is joined with microwave, ultrasound, and subcritical water to hydrolyze fucoidan (Alboofetileh et al. 2018). In the subcritical water process, 0.1% NaOH has been shown to be more efficient in extracting fucoidan from brown seaweed compared to pure water (Zhang et al. 2024). The increased yield with alkalinity may be due to the disruption of chemical bonds between solute and solid, facilitating the release of fucoidan into the solvent (Poojary et al. 2022). Optimal conditions for the isolation of alginate and fucoidan from "waste" beach-cast brown algae with high yields were found to be at 160 °C, 18 min, and a solid mass of 0.1 g. However, high temperatures at 200 °C improved phenolic yield and antioxidant activities but led to degradation of alginate and fucoidan (Poojary et al. 2022). Furthermore, a combination of EAE and UAE is utilized to hydrolyze phenolic constituents and monosaccharides like glucose, arabinose, fucose, and mannose (Casas et al. 2019). A combination of EAE and MAE methods is used to extract phlorotannin from seaweed. The use of this combination provides high performance and a shorter time compared to enzymatic and conventional methods (Charoensiddhi et al. 2015).

4.3.3 *Pressurized Liquid Extraction (PLE)*

Pressurized liquid extraction (PLE), also known as accelerated solvent extraction (ASE) or pressurized solvent extraction (PSE) is a sustainable technique that involves using high temperatures between 20–50 °C and a pressure of 10–15 MPa to separate components from a solid matrix. This method uses low quantities of solvent to reduce the negative impacts of toxic solvents on human health and the environment, potentially improving the overall sustainability of the process, as shown in Fig. 4.2. PLE makes it possible to use eco-friendly alternatives such as water or a combination of water with eutectic solvents or ionic liquids (Ummat et al. 2021). The process is also referred to as subcritical, superheated, or pressurized hot-water extraction when water is utilized as the solvent (Essien et al. 2020).

Please provide high-resolution source file for Figs. 4.2 and 4.3. Modified and send as attached file.

Additionally, PLE doesn't need extra processes for separation or concentration before analysis (Poojary et al. 2022). Furthermore, PLE requires less extraction time (5 to 20 min) than other conventional extraction methods, such as Soxhlet extraction (Ballesteros-Vivas et al. 2020). Although it is less selective than SFE and unsuitable

High temperatures enable a reduction in solvent viscosity, which, as mentioned earlier, speeds up the target compound's solubilisation and diffusion (Mena-García et al. 2019). According to Pavkovich and Bell (2019), extraction typically takes 15 min, which lowers the energy needed for the operation and makes this technology more environmentally friendly than other traditional methods.

4.3.3.1 Sample Pre-Treatments

To enhance the contact surface between the solvent and the matrix during extraction, samples must be pre-treated before using PLE (Carabias-Martínez et al. 2005).

The four steps of pre-treatment are as follows, as illustrated in Fig. 4.3:

1. Drying: drying is the first step in improving extraction efficiency, with the goal of removing water from the matrix (Carabias-Martínez et al. 2005). Lyophilization is the preferred technique as it preserves the chemicals and is faster than oven heating or air drying. When using non-polar solvents, it is important to include a desiccant in the PLE cell, a step that is often overlooked (Nieto et al. 2010).
2. Homogenization: The goal of homogenization is to uniformly disperse the sample through grinding (Nieto et al. 2010).

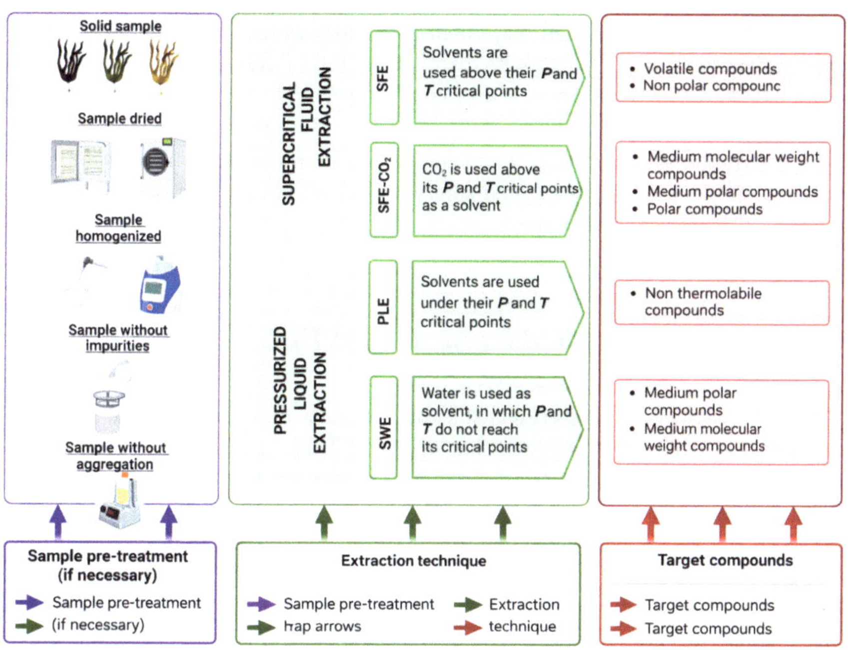

Fig. 4.3 A schematic representation of the pre-treatment steps, extraction methods and target compounds from seaweeds using PLE and SWE (Perez-Vazquez et al. 2023)

3. Sieving: the process of sieving the analyte increases its surface area and facilitates its diffusion from the matrix to the solvent (Carabias-Martínez et al. 2005). For PLE, a 2 mm sieve is frequently utilized, and this step usually yields a similar particle size (Nieto et al. 2010). Following sieving, the separated fraction can be ground into larger particles for additional processing.
4. For specific samples, it is advised to disperse them with an inert agent to avoid particle agglomeration, which could alter the efficacy of the extraction (Carabias-Martínez et al. 2005).

This process keeps the water in a liquid state by heating it above its typical boiling point (50–300 ºC) but below the critical temperature and pressure (35–200 bar) (Matos et al. 2021). While alternative solvents can be used, water is the recommended solvent since it prevents the production of harmful waste. As stated by Khalil et al. (2017), water has a special behavior whereby its dielectric constant decreases with temperature, thereby enhancing its affinity for less polar molecules. A general improvement in extraction kinetics is also possible due to the combination of high temperature and pressure, which also makes it easier for chemicals to desorb from the material, enhances solvent solubility, and encourages diffusion into materials. Subcritical water extraction (SWE) has a number of advantages, including non-hazardous circumstances, faster extraction periods, low solvent use, and excellent extraction efficiency (Bordoloi and Goosen 2020). In this regard, Saravana et al. (2018) extracted 13% of the fucoidans from *Saccharina japonica* using SWE. Protein fractions of red algae *Gelidium sesquipedale* residues were subjected to SWE procedures by Trigueros et al. (2021), who demonstrated its effectiveness. The study made it clear that the severity variables of the extraction process are directly correlated with non-polar selectivity. The primary drawback of SWE is the high expense of maintenance and equipment scaling. The necessity of high operating temperatures and pressures, along with the associated safety measures, are the main causes of these expenses (Gomes-dias et al. 2020).

4.3.4 *Supercritical Fluid Extraction (SFE)*

A fluid that reaches the supercritical state has a density that similar to that of a liquid and a viscosity that falls between that of a liquid and a gas. This allows solutes to be more easily transferred from the matrix to the solvent (Bordoloi and Goosen 2020). CO_2 is the most widely used solvent due to its critical point (Tc: 31 ºC, Pc: 7.38 MPa), which enables the extraction of labile compounds. Furthermore, CO_2 is readily available, non-explosive, non-toxic, and non-flammable (Chemat et al. 2017). The main advantage of using CO_2 is the lack of residue from the extraction process during decompression. Furthermore, the extraction of oxidation-prone chemicals is supported by CO_2's non-oxidative character (Essien et al. 2020). A disadvantage of applying CO_2 for SFE is that it can only extract non-polar molecules because of its chemical behavior, which is comparable to that of lipophilic or non-polar solvents. In order to get around this restriction, polar solvents, including ethanol, methanol, and water, are employed as co-solvents to alter the solvent's polarity (Molino et al. 2020). Although ethanol, ethane, methanol, methane, propylene, propane, ethylene, and acetone are additional possible supercritical solvents, none of them fit the definition of "green chemistry and engineering" (Bordoloi and Goosen 2020). The key factors influencing the extraction process in this technology include the type of fluid, pressure (100–400 bar), temperature (40–90 °C), moisture content, solvent shape, flow rate, and material size.

Significant advantages often include non-flammability, low cost, availability of solvents, minimal time consumption, effective extraction, and high selectivity (Khalil et al. 2017).

Antioxidants and neutral lipids are the most often recovered chemicals from algae using the SFE method because CO_2 works well as a solvent for non-polar target molecules, while an organic co-solvent, like ethanol, is used for more polar molecules (Khalil et al. 2018). The food and pharmaceutical industries use the low polar seaweed biocompounds extracted with SFE, such as polyphenols, fatty acids, and fucoxanthin (Bogolitsyn et al. 2017; Bordoloi and Goosen 2020). SFE technology was used by Bogolitsyn et al. (2017) to produce 20 mg/g of fatty acids and polyphenols from brown seaweeds (*Fucus* sp.), while Roh et al. (2008) used a cosolvent to obtain approximately 49 mg/g of carotenoids from *Undaria* sp. Further study is necessary to maintain the practicality of scaling up SFE, as its energy consumption and high equipment expense make it the most expensive option for mass production (Bordoloi and Goosen 2020).

4.3.5 Electric-Field-Based Technology

Electric fields or electromagnetic waves in electro-technologies are considered the best techniques to remove bioactive substances from algal biomass (Pereira et al. 2023). Brief pulses of extremely high voltage electric fields cause electroporation, an environmentally friendly technique that increases the permeability of cell membranes. Electric fields and temperature can work together to make electric effects extractions beneficial and open up new opportunities for the processing business. These electric technologies work by passing an electric current through biomaterials, reducing the need for solvents, improving product quality, increasing efficiency, and lowering operating and energy expenses (Rocha et al. 2018). Various events may occur during the administration of electric fields, depending on their intensity, current flow style (direct or alternating), electric frequency, and treatment period. These events include: (i) electroporation, in which an external electric field (between 0.2 and 1 V) causes changes in the transmembrane potential, microorganism inactivation, and cellular material.

These effects depend on the exposure period, electric field intensity, and medium composition. (ii) Electrophoresis is a charge-related process that may occur when molecules are subjected to a continuous electric field current; (iii) Ohmic heating (OH) occurs when the system works as a semiconductor, leading in internal heat dissipation explained by the Joule effect; and (iv) Electrical frequency, electric field intensity, treatment duration, and the characteristics of treated solutions (such as pH, ionic strength, and electrical conductivity) all impact electrochemical processes. Additionally, secondary phenomena such as the creation of shock waves, electrochemical reactions, and light emission may occur. The fundamental technologies of electric field processing fall into the following categories: (i) Pulsed electric fields (PEF) and high voltage electrical discharge (HVED) are examples of "nonthermal" pulsed electrical applications that apply high intensity for microseconds or nanoseconds, typically between 10 and 60 kV/cm; (ii) non-pulsed electric applications, which employ sinusoidal or quadratic waves with varying frequencies (from 50 Hz to kHz) and unrestricted processing times to apply moderate electric fields (< 1 kV/cm).; and (iii) dielectric heating technologies like microwave-assisted extraction (MAE) and radiofrequency (RF), which are based on the application of high-frequency electromagnetic radiation that results in heat production due to the oscillation of dipolar molecules (Pereira et al. 2023; Rocha et al. 2018).

Regardless of this classification, it is key to emphasize that OH is always a side consequence of electrical current flowing through a material and can be appropriately managed based on how electric waves are transmitted. Because of this, published research also emphasizes therapies like Pulsed Ohmic Heating (POH), which is based on the use of electrical pulses intended to dissipate heat via the OH effect. Additionally, these technologies can be applied as a pretreatment to break up the structure of the algae prior to a traditional or alternative extraction process. The key strengths and weaknesses of each electric field-based technology are shown in Table 4.4.

4.3.5.1 Microwave-Assisted Extraction (MAE)

MAE most widely utilized, reasonably priced, and non-chemical technology in laboratories. Using microwave potential (often 2.45 GHz) with polar solvent molecules forms the basis of this technique, which causes changes in cell structure due to electromagnetic microwaves. This electromagnetic energy is converted into heat energy through ionic conduction and dipole rotation (Mena-García et al. 2019). In other words, MAE functions inside the nonionizing electromagnetic wave spectrum, which ranges from 300 MHz to 300 GHz. Depending on the temperature, the absorption of these waves can result in two distinct processes: ionic conduction and dipole rotation, which alter the structural makeup of the matrix. The basis for dipole rotation is the alignment of the samples' dipolar fractions with the microwaves' electric field, which causes them to collide and produce heat. The basis for ionic conduction is the movement of charged ions as a result of the microwaves' influence on the electrostatic field; with heat being produced by the friction of these movements (Khalil et al. 2018; Zollmann et al. 2019). Frequency, power, solid-to-liquid ratio, solvent concentration, solvent properties, matrix properties, temperature, number of extraction cycles, extraction duration and pressure are all crucial variables to take into account (Matos et al. 2021). This extraction method is faster than conventional extraction in producing

Table 4.4 Main frailties and advantages of electric field-based technologies (Pereira et al. 2023)

Electric-field technology	Frailty	Advantage	References
Microwave-assisted extraction	Scale-up process due to limitation of waves penetration into the feedstock matrices and high equipment and operation costs	Efficiency, cost-effective, non-chemical, consistency, selective, time-saving, less energy and solvent consumption	Abdul Khalil et al. (2018)
Pulsed electric fields	Scale-up equipment at an industrial level, parameter standardization, electrodes oxidation, fouling, and bubbling	Efficiency, cost-effective, time-saving, significant disruption effect, less energy and solvent consumption, non-chemical	Matos et al. (2021)
High voltage electrical discharge	Non-selective, aggressiveness, operational volume, process control, industrial implementation	High efficiency, high levels of cellular damage or disintegration, much-reduced energy input, time-saving, less solvent consumption, non-chemical	Boussetta and Vorobiev (2014)
Moderate electric fields and Ohmic heating	Electrode corrosion	Versatility, scale-up easiness, efficiency, cost-effective, time-saving, less energy and solvent consumption, homogeneous heating, selective, high quality products non-chemical	Pereira et al. (2016)
Pulsed ohmic heating	Process control, industrial scale-up	Efficiency, less energy and solvent consumption, time-saving, cost-effective, non-chemical	Pereira et al. (2023)

bioactive compounds (Vinatoru et al. 2017). In addition to extracting diverse biocompounds like polysaccharides and phenolics from various seaweed groups, MAE can be applied directly to fresh biomass. Consistency and efficiency, the capacity to heat raw materials locally and selectively, enhanced mass transfer and tissue disruption, reduced energy and solvent consumption, cost-effectiveness, high extraction rates, and high-quality products are just a few of the many benefits of MAE.

MAE systems are divided into two categories: closed and open. The open system eliminates the possibility of an explosion, is completely automatic, and is less expensive. It also operates under air pressure. However, compared to closed systems, which run at higher pressure and temperature and have an explosion danger, it is less accurate, cannot work with multiple samples at once, and takes longer extraction periods (Grosso et al. 2015; Khalil et al. 2018). Although there are significant limitations because of the restricted wave penetration into the feedstock matrices, MAE can be scaled up to the industrial level. Further research is necessary because of the high equipment costs and operator skills required, as well as the substantial initial cash investment needed (Bordoloi and Goosen 2020; Khalil et al. 2018). However, it is very difficult to extract thermo labile chemicals like proteins, pigments, and fatty acids because of the technology's rapid heat generation and the difficulty in precisely managing temperature (Bordoloi and Goosen 2020). Magnusson et al. (2017) demonstrated that, in comparison to traditional extraction techniques, MAE might enhance *Carpophyllum flexuosum* polyphenol yields by up to 70%.

4.3.5.2 Pulsed Electric Field (PEF)

A non-thermal technique called PEF uses extremely high voltage pulses (1–100 kV/cm) for brief periods of time (nano or microseconds) to cause the cell membrane to electropermeate (Pereira et al. 2023). The intensity of the electric field used determines the efficiency of this extraction process; values below 25 kV/cm are associated with compound extraction, while values above 25 kV/cm are used for microbial inactivation (Pereira et al. 2023). Promising and alluring electro-technologies for extractions include ohmic heating (OH) and pulsed electric field (PEF) (Pereira et al. 2016). This technology can be used to extract various chemicals (e.g., proteins, carbohydrates, etc.) and create stress in a wide range of target applications, including microalgae, bacteria, plant tissues, seaweeds, etc. In addition to its adaptability, PEF treatment is highly effective, uses little water and energy, produces little heat, doesn't require chemicals, and is appropriate for industrial settings (Pereira et al. 2016; Zollmann et al. 2019). PEF technology has several disadvantages, including high generator costs, difficulties in controlling temperature, and issues such as electrode oxidation, erosion, fouling, and bubbling (Pereira et al. 2016).

The literature offers a variety of PEF techniques that combine different parameters such as temperature, wave shape, electrode geometry and material types, specific energy, treatment time, electric field strength, number and duration of pulses, treatment chamber design, and properties of the product and media. Despite the availability of industrial PEF applications, equipment scale-up and proper control and standardization of these parameters are still very challenging (Pereira et al. 2023; Rocha et al. 2018). A high disruptive effect is expected with PEF despite its limited selectivity, making it an attractive option to improve extraction yields from algae and overcome the problems of cell rupture that some species of algae face.

4.3.5.3 High Voltage Electrical Discharge (HVED)

HVED causes a sudden release of high energy by applying high-current electric pulses directly into a liquid medium. This leads to the formation of a plasma channel and a number of secondary phenomena, including shock waves, radical generation, cavitation, heat dissipation, light emission, and electroporation (Boussetta and Vorobiev 2014). Since it allows for high levels of cellular damage or disintegration comparable to traditional mechanical processing techniques, but with the benefit of a far lower energy input, this technology has become an appealing alternative. However, the primary disadvantages of HVED that prevent its industrial application are its non-selective release of all cellular material, operational volume limitation, procedure aggressiveness, and lack of process control (Boussetta and Vorobiev 2014). According to Grimi et al. (2014), despite its disadvantage, the electrically based disruption technique HVED improves the selective recovery of water-soluble proteins, water-soluble ionic components, small molecular weight organic compounds, and microelements from the microalga *Nannochloropsis* sp. However, it was found to be ineffective for extraction of pigment.

4.3.5.4 Moderate Electric Fields and Ohmic Heating Effects

The term "moderate electric fields" (MEF) refers to an application of electric fields that uses low electrical intensities (typically between 1 and 1000 V/cm), longer electrical treatment times, a frequency range of 50 Hz to a few MHz, and a defined waveform (usually square or sinusoidal) (Pereira et al. 2023; Rocha et al. 2018). Ohmic Heating (OH) is an effect that occurs when MEF is applied to a semi-conductive material. According to Joule's effect, heat dissipation is caused by a conductivity-dependent mechanism. This technique offers rapid, uniform heating with high energy efficiency, which promotes permeability in cell membranes for effective drug extraction. It is also used to activate or deactivate enzymes. Because of its adaptability and ease of scaling up, the OH/MEF combination processing has been successfully used in extraction operations in the industry. While MEF has received less attention in the literature than PEF or MAE, the data that has been published suggests that it has enormous promise, for processes that operate in mild settings and situations where temperature plays a significant role. The use of MEF and mild OH effectively enhanced the extraction of bioactive compounds, specifically phycobiliproteins like C-phycocyanin, a significant natural blue pigment, from the cyanobacteria *Arthrospira platensis* at temperatures between 30 and 51 °C (Ferreira-Santos et al. 2020). They also observed increased resistance to thermal degradation of the pigment and a cell disruption impact in the MEF. The extraction of other pertinent biomolecules from algae, such as proteins, polysaccharides, pigments (including carotenoids), phenolic compounds, and lipids, was successful when MEF with regulated OH was utilized. Another noteworthy aspect documented aspect is that, while MEF alone does not increase process selectivity, it appears to increase the solvent's selective power, which enhances both process efficiency and selectivity (Pereira et al. 2023).

Conversely, a sample is heated using a moderate electric field (MEF) in Ohmic heating (OH). This is a green technology that efficiently transfers energy efficiently and offers consistent, quick, and accurate heating. The rapid heating rate and uniform, instantaneous thermal energy transfer enable high-temperature applications in short intervals, improving cell membrane permeability and yielding high-quality products with compounds preserved, in contrast to traditional extraction methods (Pereira et al. 2016; Aamir and Jittanit 2017). In accordance with Joule's Law, heat is produced when electric current flows through materials that are to be heated (such as a mixture of seaweed and solvent) that are in contact with the electrodes. This green technology depends on a number of parameters, including temperature, electric field intensity (ranging from 1 to 1000 V/cm, which defines a moderate electric field), frequency range (50 to 25.000 Hz), type, length of treatment, and electrical conductivity (ranging from 0.01 to 10 S/m). Other benefits of OH that have been mentioned include high yields, quick extraction times, reduced nutritional value loss, ease of scale-up, and preservation of color pigments. However, there are few instances of the OH technique being used with seaweeds (Matos et al. 2021). As a pretreatment to break down biomass and make extraction easier with other complementing technologies, PEF is seen by some writers as having good potential in integrated seaweed biorefinery systems (Zollmann et al. 2019). In their 2015 study, Postma et al. examined the effects of PEF on water-soluble proteins and carbohydrates of *Ulva lactuca*, obtaining yields of less than 15% and 15%, respectively. Using Pulsed Ohmic Heating (POH), Robin et al. (2018) demonstrated a 20% protein production from the same algal species. Similar to PEF, the primary disadvantages of this green process are the potential for electrolysis, electrode degradation, and metal leakage into the medium due to low frequency operation (Pereira et al. 2016; Aamir and Jittanit 2017). When PEF is applied, the combined actions of OH and PEF can be used to improve cell permeabilization and heat dissipation. Though they are often ignored or not accurately measured, it is crucial to understand that OH effects may naturally occur during PEF application. In contrast to MEF, PEF requires additional energy input, process management, and industria scale-up, but it may increase treatment efficiency due to its capacity to dissipate a sizable and manageable amount of heat (Pereira et al. 2023).

4.3.6 The Combination of Different Green Methods

Regarding strategies for valorization, algae constitute a very diversified biomass with numerous options. By combining many process types and attempting to utilize each technology's greatest qualities in order to recover distinct target fractions from various biomass structures, while also getting around some of the drawbacks of each particular technique. The combination of alternative technologies can take place concurrently or consecutively with the aim of improving cell wall permeability and cell disruption, making the extraction process easier. This approach may involve combination of physical and chemical effects, such as temperature, mechanical damage, and electrochemical reactions, to achieve high selectivity for desired compounds, energy efficiency, and extraction effectiveness.

4.3.6.1 Combinatorial Approaches of PLE with Other Extraction Methodologies

There have been few studies on the use of PLE in seaweeds due to its novelty. However, following green technology trends, PLE could be combined with other methodologies to improve bioactive compound extraction efficiency and reduce solvent and time consumption. Moreover, because combinatorial approaches of PLE applied in the extraction of bioactive compounds of seaweeds have not been thoroughly studied, results of trials where PLE and other techniques are applied in other matrices are shown in this section and compiled in Fig. 4.3 as case studies for future applications of these combined techniques.

PLE Combined with SPE

The combination of PLE and SPE was used to separate certain phenolic chemicals. This technology utilizes PLE's ability to extract bioactive chemicals from the matrix and SPE's ability to purify the extracted molecules (Souza et al. 2021). Because PLE is non-selective, SPE is typically utilized after extraction. While there is no evidence exists on the use of PLE and SPE to extract phenolic chemicals from seaweeds, it has been used successfully in other matrices such as mate leaves, and apple pomade. In these matrices, greater yields of total flavonoids were produced than utilizing PLE alone (Chaves et al. 2022).

PLE Combined with UAE

Phenolic chemicals were extracted from three brown algae and one red algae using PLE and UAE. The operational conditions involved using a solvent of 80% MeOH:H_2O (v/v) in a 10 mL volume at 130 °C; and 130 bar pressure under two static cycles of 10 min. These conditions increased the phenolic yield extracted from the matrix because of the high and prolonged pressure (Klejdus et al. 2017). Nevertheless, more investigation is required to ascertain whether scaling up is financially viable, as many businesses encounter difficulties because of the extraction time, high costs of facilities, and ultrasonic power (Dias et al. 2021).

PLE Combined with SFE-CO_2

It appears that the PLE-SFE-CO_2 combination is the best option because of the higher yields obtained and the profitability of scaling up. In contrast to the SFE method, PLE uses a wider variety of solvents and functions without approaching the liquid solvent's critical point (Kadam et al. 2015). The sequential procedure of combining PLE and SFE-CO_2 based on SFE-CO_2's capacity to extract the matrix's lipophilic portion and PLE's capacity to extract highly polar or antioxidant compounds (Kraujalis et al. 2017). Furthermore, an economic assessment of the procedure applied to by-products of passion fruit was conducted in Brazil. There is no information available regarding this sequential method employed in seaweeds, but this study shows that the combination of these strategies is economically suitable for large-scale production. It boosts process productivity and lowers manufacturing costs (Viganó et al. 2017).

PLE Combined with EAE

As previously mentioned, the use of EAE as an extraction approach has several drawbacks, such as the expensive cost of enzymes, restrictions on cell disruption caused by enzyme specificity, and enzyme deactivation from pH and temperature variations (Das et al. 2021). To overcome these constraints, research has integrated various approaches. For example, EAE was used in conjunction with PLE and alkaline hydrolysis to extract bioactive chemicals from *S. muticum*. The extraction yields were greater than when PLE was used alone. The development of a protein–polyphenol complex, which causes aggregation and precipitation, eventually leading to decreased yields, may help to explain this. Proteins and other substances may be released as the enzyme breaks down the seaweed's cell, forming this complex (Otero et al. 2021). It was reported by Steinbruch et al. (2023) that PEF and EAE were sequentially combined. PEF treatment was followed by enzymatic cell wall destruction from *Ulva* sp. to produce a protein extract with strong antioxidant activity. Further research is needed to determine if combining EAE and SPE could result in higher yields.

4.3.6.2 Other Combined Strategies Using Electric Field-Based Technologies

Combining the fast-heating rates of SWE and MAE has the potential to increase water affinity for less polar molecules by decreasing the dielectric constant. Heat transfer in MAE occurs quickly because of volumetric heating. The discovery of a synergistic relationship between saccharide and lipid and an antagonistic interaction between protein and lipid under microwave irradiation, has led to several studies that combined MAE and SWE simultaneously as a promising pathway for the generation of biocrude. Even though subcritical water has unique properties, other subcritical solvents can be used with MAE to take advantage of higher heating rates, energy efficiency, and improved process kinetics, mass transfer rates, and solubility at higher temperatures. The primary goal of PEF as a pretreatment is to disrupt the structure of the algae. Therefore, it can be used in conjunction with any other suitable extraction method later on. For example, Guo et al. (2019) found that the sequential combination of PEF and SWE improves the extraction of valuable compounds, such as lipids and proteins, from microalgae biomass. This is achieved first by PEF pretreatment and then by SWE of the residue to increase the production of high quality biocrude. In some instances, the presence of an electric field may alter enzyme activity, protein functionality, and molecular interactions. In conjunction with EAE, MEF effects can be adjusted to improve extraction efficiency and selectivity. Additionally, applying mild electric fields may cause the enzymes' structures to change, increasing their activity (Ferreira-Santos et al. 2020). In order to recover agar from red seaweeds, a recent study combined SWE and MEF simultaneously. It shown that moderate electric fields increased agar extraction yields by up to 40% while also improving processing time and efficiency (Pereira et al. 2023). Depending on the technologies' propensity to recover distinct target fractions, the process may be set up to employ them successively. McElroy et al. (2023) used the sequential combination of SFE and MAE to evaluate the viability of this novel method for saccharina latissima valorisation. SFE selective power towards non-polar compounds was firstly used to recover fucoxanthin (that is also thermolabile), while MAE or conventional protocols were further used to recover mannitol, alginates and fucoidans. The results showed that combining supercritical carbon extraction with MAE would increase the recovery of multiple valuable compounds with bioactive properties from algae biomass.

4.3.7 Fermentation

Fermentation has emerged as an alternative to standard methods for extracting high-quality bioactive chemicals without the use of hazardous solvents. It is the process through which carbon molecules such as glucose and other polysaccharides are transformed into smaller molecules with lower molecular weights, such as acetic, lactic, and other organic acids. This molecular oxidation/reduction by specific microorganisms (Yadav et al. 2012). Fermentation of seaweed biomass is a relatively new field, but it shows great promise for the development of new food products that retain and improve the organoleptic and beneficial qualities that these organisms are known for. Although more research is needed in this area, fermented seaweed's prebiotic and probiotic potential can help accelerate the creation of new functional foods. In this circumstance, seaweed might provide solutions for various industries. Fermentation has the potential to lower expenses and spur creativity in the production of food and animal feed (Monteiro et al. 2021).

According to research, bioactive chemicals found in seaweeds can be used as a substrate for lactic acid bacteria that specialize in lactic acid synthesis (Bruhn et al. 2019). This colorless and odorless monocarboxylic acid has been positively tested for use in food, pharmaceuticals, and industry (Mora-Villalobos et al. 2020). The inclusion of seaweeds as a prebiotic ingredient could potentially address cost concerns. Additionally, the consumption of non-digestible seaweed polysaccharides may promote the growth of lactic acid bacteria in our intestines. This, in turn, supports the proliferation of beneficial bacteria while inhibiting pathogenic bacteria (Monteiro et al. 2021).

Seaweed fermentation is not well-documented in literature, for instance, the fermentation of *Undaria pinnatifida* (Phaeophyceae) has been suggested as a different approach for feeding aquaculture, and a fermented beverage made from

Gracilaria fisheri (Rhodophyta) has been created for human consumption within the study conducted by Martelli et al. (2020).

4.4 Conclusion

Recently, the recovery of biocompounds from marine organisms such as seaweeds and microalgae has become a hot topic, with many industries seeking to develop functional products from these marine organisms. In this comprehensive chapter, we reviewed research articles mainly focused on the conventional and modern methods used for extracting biocompounds from seaweed. These methods were used while taking into account the recovery rate, time, volume, temperature, pH, ratio, and kind of solvent, as well as the procedure. An important consideration in determining the environmentally friendly extraction techniques to use is the bioactive target to be extracted from seaweed. This ensures that the bioactive compound obtained is maximal, effective and efficient. These different techniques have the potential to significantly contribute to the development of new functional products from algae in various industries. However, further research studies are needed to optimize conditions and identify the most effective combination of these techniques for specific compounds.

References

Aamir M, Jittanit W (2017) Ohmic heating treatment for Gac aril oil extraction: effects on extraction efficiency, physical properties and some bioactive compounds. Innov Food Sci Emerg Technol 41:224–234

Alboofetileh M, Rezaei M, Tabarsa M, Ritta M, Donalisio M, Mariatti F, You SG, Lembo D, Cravotto G (2018) Effect of different non-conventional extraction methods on the antibacterial and antiviral activity of fucoidans extracted from *Nizamuddinia zanardinii*. Int J Biol Macromol 124:131–137

Ballesteros-Vivas D, Alvarez-Rivera G, Ibánez E, Parada-Alfonso F, Cifuentes A (2019) Integrated strategy for the extraction and profiling of bioactive metabolites from Passiflora mollissima seeds combining pressurized-liquid extraction and gas/liquid chromatography–high resolution mass spectrometry. J Chromatogr A 1595:144–157

Ballesteros-Vivas D, Ortega-Barbosa JP, del Pilar Sanchez-Camargo A, Rodríguez-Varela LI, Parada-Alfonso F (2020) Pressurized liquid extraction of bioactives. In: Comprehensive foodomics. Elsevier, Amsterdam, The Netherlands, pp 754–770. ISBN 9780128163955

Bogolitsyn KG, Kaplitsin PA, Dobrodeeva LK, Druzhinina AS, Ovchinnikov DV, Parshina AE, Shulgina EV (2017) Fatty acid composition and biological activity of supercritical extracts from arctic brown algae *Fucus vesiculosus*. Russ J Phys Chem B 11:1144–1152

Bordoloi A, Goosen N (2020) Green and integrated processing approaches for the recovery of high-value compounds from brown seaweeds, vol 95. Elsevier, Amsterdam, The Netherlands. ISBN 9780081027103

Boussetta N, Vorobiev E (2014) Extraction of valuable biocompounds assisted by highvoltage electrical discharges: a review. C R Chim 17:197–203. https://doi.org/10.1016/j.crci.2013.11.011

Bruhn A, Brynning G, Johansen A, Lindegaard MS, Sveigaard HH, Aarup B, Fonager L, Andersen LL, Rasmussen MB, Larsen MM et al (2019) Fermentation of sugar kelp (*Saccharina latissima*)—effects on sensory properties, and content of minerals and metals. J Appl Phycol 31:3175–3187

Budhiraja RP (2004) Separation chemistry. New Age International Publisher, New Delhi

Calvo-Flores FG (2020) Green processes in foodomics. Green solvents for sustainable processes. In: Comprehensive foodomics. Elsevier, Amsterdam, The Netherlands, pp 690–709. ISBN 9780128163955

Carabias-Martínez R, Rodríguez-Gonzalo E, Revilla-Ruiz P, Hernández-Méndez J (2005) Pressurized liquid extraction in the analysis of food and biological samples. J Chromatogr A 1089:1–17

Carreira-Casais A, Otero P, Garcia-Perez P, Garcia-Oliveira P, Pereira AG, Carpena M, Soria-Lopez A, Simal-Gandara J, Prieto MA (2021) Benefits and drawbacks of ultrasound-assisted extraction for the recovery of bioactive compounds from marine algae. Int J Environ Res Public Health 18:9153

Casas MP, Conde E, Domínguez H, Moure A (2019) Ecofriendly extraction of bioactive fractions from Sargassum muticum. Process Biochem 79:166–173

Charoensiddhi S, Franco C, Su P, Zhang W (2015) Improved antioxidant activities of brown seaweed *Ecklonia radiata* extracts prepared by microwave-assisted enzymatic extraction. J Appl Phycol 27(5):2049–2058

Chaves JO, Sanches VL, Viganó J, de Souza Mesquita LM, de Souza MC, da Silva LC, Acunha T, Faccioli LH, Rostagno MA (2022) Integration of pressurized liquid extraction and in-line solid-phase extraction to simultaneously extract and concentrate phenolic compounds from lemon peel (Citrus limon L.). Food Res Int 157:111252

Chemat F, Rombaut N, Meullemiestre A, Turk M, Perino S, Fabiano-Tixier AS, Abert-Vian M (2017) Review of green food processing techniques, preservation, transformation and extraction. Innov Food Sci Emerg Technol 41:357–377

Dang TT, Bowyer MC, Van Altena IA, Scarlett CJ (2017a) Comparison of chemical profile and antioxidant properties of the brown algae. Int J Food Sci Technol 53(1):174–181

Dang TT, Van Vuong Q, Schreider MJ, Bowyer MC, Altena IAV, Scarlett CJ (2017b) Optimisation of ultrasound-assisted extraction conditions for phenolic content and antioxidant activities of the alga *Hormosira banksii* using response surface methodology. J Appl Phycol 29:3161–3173

Das S, Nadar SS, Rathod VK (2021) Integrated strategies for enzyme assisted extraction of bioactive molecules: a review. Int J Biol Macromol 191:899–917

Daud NM, Putra NR, Jamaludin R, Norodin NSM, Sarkawi NS, Hamzah MHS, Nasir HM, Zaidel DNA, Yunus MAC, Salleh LM (2022) Valorisation of plant seed as natural bioactive compounds by various extraction methods: a review. Trends Food Sci Technol 119:201–214

Dias ALB, de Aguiar AC, Rostagno MA (2021) Extraction of natural products using supercritical fluids and pressurized liquids assisted by ultrasound: current status and trends. Ultrason Sonochem 74:105584

Dulanlebit YH, Hernani (2023) Overview of extraction methods for extracting seaweed and its applications IPA. J Res Sci Educ

Essien SO, Young B, Baroutian S (2020) Recent advances in subcritical water and supercritical carbon dioxide extraction of bioactive compounds from plant materials. Trends Food Sci Technol 97:156–169

Fabrowska J, Messyasz B, Szyling J, Walkowiak J, Łęska B (2018) Isolation of chlorophylls and carotenoids from freshwater algae using different extraction methods. Phycol Res 66(1):52–57

Ferreira-Santos P, Nunes R, Biasio FD, Spigno G, Gorgoglione D, Teixeira JA, Rocha CMR (2020) Influence of thermal and electrical effects of ohmic heating on C-phycocyanin properties and biocompounds recovery from Spirulina platensis. LWT Food Sci Technol 128:109491. https://doi.org/10.1016/j.lwt.2020.109491

Fleurence J, Massiani L, Guyader O, Mabeau S (1995) Use of enzymatic cell wall degradation for improvement of protein extraction from *Chondrus crispus*, *Gracilaria verrucosa* and *Palmaria palmata*. J Appl Phycol 7:393–397

Garcia-Vaquero M, Rajauria G, Tiwari B (2020) Conventional extraction techniques: solvent extraction. In: Sustainable seaweed technologies. Elsevier, Amsterdam, The Netherlands, pp 171–189

Grosso C, Valentão P, Ferreres F, Andrade PB (2015) Alternative and efficient extraction methods for marine-derived compounds. Mar Drugs 13(5):3182–3230

Gomes-dias JS, Roman A, Teixeira JA, Rocha CMR (2020) Valorization of seaweed carbohydrates: autohydrolysis as a selective and sustainable pretreatment. ACS Sustain Chem Eng 8:17143–17153

Grimi N, Dubois A, Marchal L, Jubeau S, Lebovka NI, Vorobiev E (2014) Selective extraction from microalgae Nannochloropsis sp. using different methods of cell disruption. Bioresour Technol 153:254–259. https://doi.org/10.1016/j.biortech.2013.12.011

Guo B, Yang B, Silve A, Akaberi S, Scherer D, Papachristou I, Frey W, Hornung U, Dahmen N (2019) Hydrothermal liquefaction of residual microalgae biomass after pulsed electric field-assisted valuables extraction. Algal Res 43:101650

Habeebullah SF, Alagarsamy S, Sattari Z, Al-Haddad S, Fakhraldeen S, Al-Ghunaim A, Faiza A-Y (2020) Enzyme-assisted extraction of bioactive compounds from brown seaweeds and characterization. J Appl Phycol 32:615–629

Hardouin K, Bedoux G, Burlot A-S, Collén PN, Bourgougnon N (2014a) Enzymatic recovery of metabolites from seaweeds: potential applications. Adv Bot Res 71:279–320

Hardouin K, Burlot A-S, Umami A, Tanniou A, Stiger-Pouvreau V, Widowati I, Bedoux G, Bourgougnon N (2014b) Biochemical and antiviral activities of enzymatic hydrolysates from different invasive French seaweeds. J Appl Phycol 26:1029–1042

Hardouin K, Bedoux G, Burlot A, Donnay-moreno C, Bergé J, Nyvall-collén P, Bourgougnon N (2016) Enzyme-assisted extraction (EAE) for the production of antiviral and antioxidant extracts from the green seaweed *Ulva armoricana*. ALGAL 16:233–239

Jacobsen C, Sørensen ADM, Holdt SL, Akoh CC, Hermund DB (2019) Source, extraction, characterization, and applications of novel antioxidants from seaweed. Annu Rev Food Sci Technol 10:541–568. https://doi.org/10.1146/annurev-food-032818-121401

Ju ZY, Howard LR (2005) Subcritical water and sulfured water extraction of anthocyanins and other phenolics from dried red grape skin. J Food Sci 70(4):270–276

Kadam SU, Tiwari BK, Smyth TJ, O'Donnell CP (2015) Optimization of ultrasound assisted extraction of bioactive components from brown seaweed *Ascophyllum nodosum* using response surface methodology. Ultrason Sonochem 23:308–316

Khalil HPSA, Lai TK, Tye YY, Rizal S, Chong EWN, Yap SW, Hamzah AA, Fazita MRN, Paridah MT (2018) A review of extractions of seaweed hydrocolloids: properties and applications. Express Polym Lett 12:296–317

Khalil HPS, Tye YY, Saurabh CK, Leh CP, Lai TK, Chong EWN, Syakir MI (2017) Biodegradable polymer films from seaweed polysaccharides: a review on cellulose as a reinforcement material. Express Polym Lett 11(4):26. https://doi.org/10.3144/expresspolymlett.2017.26

Klejdus B, Plaza M, Šnóblová M, Lojková L (2017) Development of new efficient method for isolation of phenolics from sea algae prior to their rapid resolution liquid chromatographic–tandem mass spectrometric determination. J Pharm Biomed Anal 135:87–96

Kraujalis P, Kraujalienė V, Kazernavičiūte R, Venskutonis PR (2017) Supercritical carbon dioxide and pressurized liquid extraction of valuable ingredients from Viburnum opulus pomace and berries and evaluation of product characteristics. J Supercrit Fluids 122:99–108

Lee SH, Ko CI, Ahn G, You S, Kim JS, Heu MS, Kim J, Jee Y, Jeon YJ (2012) Molecular characteristics and anti-inflammatory activity of the fucoidan extracted from Ecklonia cava. Carbohydr Polym 89:599–606

Lim C, Yusoff S, Ng CG, Lim PE, Ching YC (2021) Bioplastic made from seaweed polysaccharides with green production methods. J Environ Chem Eng 9(5):105895. https://doi.org/10.1016/j.jece.2021.105895

Lu J, Feng X, Han Y, Xue C (2013) Optimization of subcritical fluid extraction of carotenoids and chlorophyll a from Laminaria japonica Aresch by response surface methodology. J Sci Food Agric 94(1):139–145

Magnusson M, Yuen AKL, Zhang R, Wright JT, Taylor RB, Maschmeyer T, de Nys R (2017) A comparative assessment of microwave assisted (MAE) and conventional solid-liquid (SLE) techniques for the extraction of phloroglucinol from brown seaweed. Algal Res 23:28–36

Martelli F, Favari C, Mena P, Guazzetti S, Ricci A, Del Rio D, Lazzi C, Neviani E, Bernini V (2020) Antimicrobial and fermentation potential of himanthalia elongata in food applications. Microorganisms 8:248

Matos GS, Pereira SG, Genisheva ZA, Gomes AM, Teixeira JA, Rocha CMR (2021) Advances in extraction methods to recover added-value compounds from seaweeds: sustainability and functionality. Foods 10:516. https://doi.org/10.3390/foods10030516

McElroy CR, Kopanitsa L, Helmes R, Fan J, Attard TM, Simister R, Van Den Burg S, Ladds G, Bailey DS, Gomez LD (2023) Integrated biorefinery approach to valorise Saccharina latissima biomass: combined sustainable processing to produce biologically active fucoxanthin, mannitol, fucoidans and alginates. Environ Technol Innov 29:103014. https://doi.org/10.1016/j.eti.2023.103014

Mena-García A, Ruiz-Matute AI, Soria AC, Sanz ML (2019) Green techniques for extraction of bioactive carbohydrates. TrAC Trends Anal Chem 119:115612

Molino A, Mehariya S, Di Sanzo G, Larocca V, Martino M, Leone GP, Marino T, Chianese S, Balducchi R, Musmarra D (2020) Recent developments in supercritical fluid extraction of bioactive compounds from microalgae: role of key parameters, technological achievements and challenges. J CO_2 Util 36:196–209

Monteiro P, Lomartire S, Cotas J, Pacheco D, Marques JC, Pereira L, Gonçalves AMM (2021) Seaweeds as a fermentation substrate: a challenge for the food processing industry. Processes 9:1953. https://doi.org/10.3390/pr9111953

Mora-Villalobos JA, Montero-Zamora J, Barboza N, Rojas-Garbanzo C, Usaga J, Redondo-Solano M, Schroedter L, Olszewska-Widdrat A, López-Gómez JP (2020) Multi-product lactic acid bacteria fermentations: a review. Fermentation 6:23

Muhammad M, Sayuti Mohammad S, Ketut SI, Rahmatang (2023) Environmental-friendly extraction methods to produce bioactive compounds in seaweed. Res J Chem Environ 27(11):114–121

Nieto A, Borrull F, Pocurull E, Marcé RM (2010) Pressurized liquid extraction: a useful technique to extract pharmaceuticals and personal-care products from sewage sludge. TrAC Trends Anal Chem 29:752–764

Okiyama DCG, Soares ID, Cuevas MS, Crevelin EJ, Moraes LAB, Melo MP, Oliveira AL, Rodrigues CEC (2018) Pressurized liquid extraction of flavanols and alkaloids from cocoa bean shell using ethanol as solvent. Food Res Int 114:20–29

Otero P, Carpena M, Garcia-Oliveira P, Echave J, Soria-Lopez A, Garcia-Perez P, Fraga-Corral M, Cao H, Nie S, Xiao J et al (2021) Seaweed polysaccharides: emerging extraction technologies, chemical modifications and bioactive properties. In: Critical reviews in food science and nutrition. Taylor and Francis Ltd., London, UK

Park J-S, Han J-M, Park S-W, Kim J-W, Choi M-S, Lee S-M, Haq M, Zhang W, Chun B-S (2024) Subcritical water extraction of *Undaria pinnatifida*: comparative study of the chemical properties and biological activities across different parts. Mar Drugs 22:344. https://doi.org/10.3390/md22080344

Pavkovich AM, Bell DS (2019) Extraction. In: Encyclopedia of analytical science. Elsevier, Amsterdam, The Netherlands

Pereira SG, Pereira RN, Rocha CMR, José A (2023) Teixeira Electric fields as a promising technology for the recovery of valuable biocompounds from algae: novel and sustainable approaches. Bioresour Technol Rep 22:101420
Pereira RN, Rodrigues RM, Genisheva Z, Oliveira H, de Freitas V, Teixeira JA, Vicente AA (2016) Effects of ohmic heating on extraction of food-grade phytochemicals from colored potato. LWT-Food Sci Technol 74:493–503
Perez-Vazquez A, Carpena M, Barciela P, Cassani L, Simal-Gandara J, Prieto MA (2023) Pressurized liquid extraction for the recovery of bioactive compounds from seaweeds for food industry application: a review. Antioxidants 12(3):612
Postma PR, Cerezo-Chinarro O, Akkerman RJ, Olivieri G, Wijffels RH, Brandenburg WA, Eppink MHM (2018) Biorefinery of the macroalgae *Ulva lactuca*: Extraction of proteins and carbohydrates by mild disintegration. J Appl Phycol 30:1281–1293
Poojary MM, Laurora A, Lund MN, Tiwari BK (2022) Application of pressurized liquids to extract high-value compounds from marine biomass. In: Innovative and emerging technologies in the bio-marine food sector. Elsevier, pp 441–479
Quitério E, Grosso C, Ferraz R, Delerue-Matos C, Soares C (2022) A critical comparison of the advanced extraction techniques applied to obtain health-promoting compounds from seaweeds. Mar Drugs 20(11):677
Rafiquzzaman SM, Ahmed R, Lee JM, Noh G, Jo GA, Kong IS (2016) Improved methods for isolation of carrageenan from *Hypnea musciformis* and its antioxidant activity. J Appl Phycol 28:1265–1274
Rhein-Knudsen N, Ale MT, Meyer AS (2015) Seaweed hydrocolloid production: an update on enzyme assisted extraction and modification technologies. Mar Drugs 13(6):3340–3359. https://doi.org/10.3390/md13063340
Robin A, Kazir M, Sack M, Israel A, Frey W, Mueller G, Livney YD, Golberg A (2018) Functional protein concentrates extracted from the green marine macroalga *Ulva* sp., by high voltage pulsed electric fields and mechanical press. ACS Sustain Chem Eng 6:13696–13705
Rocha CMR, Genisheva Z, Ferreira-Santos P, Rodrigues R, Vicente AA, Teixeira JA, Pereira RN (2018) Electric field-based technologies for valorization of bioresources. Bioresour Technol 254:325–339. https://doi.org/10.1016/j.biortech.2018.01.068
Roh M, Uddin S, Chun B (2008) Extraction of fucoxanthin and polyphenol from *Undaria pinnatifida* using supercritical carbon dioxide with co-solvent. Biotechnol Bioprocess Eng 2008(13):724–729
Sanagi MM, Loh SH, Wan Ibrahim WN, Pourmand N, Salisu A, Wan Ibrahim WA, Ali I (2016) Agarose-and alginate-based biopolymers for sample preparation: excellent green extraction tools for this century. J Sep Sci 39(6):1152–1159. https://doi.org/10.1002/jssc.201501207
Saravana PS, Tilahun A, Gerenew C, Tri VD, Kim NH, Kim G, Do Woo HC, Chun BS (2018) Subcritical water extraction of fucoidan from Saccharina japonica: optimization, characterization and biological studies. J Appl Phycol 30:579–590
Shang YF, Kim SM, Lee WJ, Um B (2011) Pressurized liquid method for fucoxanthin extraction from Eisenia bicyclis (Kjellman) Setchell. J Biosci Bioeng 111:237–241
Souza MC, Silva LC, Chaves JO, Salvador MP, Sanches VL, da Cunha DT, Foster Carneiro T, Rostagno MA (2021) Simultaneous extraction and separation of compounds from mate (*Ilex paraguariensis*) leaves by pressurized liquid extraction coupled with solid-phase extraction and in-line UV detection. Food Chem Mol Sci 2:100008
Steinbruch E, Wise J, Levkov K, Chemodanov A, Livney YD, Golberg A (2023) Enzymatic cell wall degradation combined with pulsed electric fields increases yields of water-soluble-protein extraction from the green marine macroalga Ulva sp. Innov Food Sci Emerg Technol 84:1–9
Suchintita DR, Tiwari BK, Chemat F, Garcia-Vaquero M (2022) Impact of ultrasound processing on alternative protein systems: protein extraction, nutritional effects and associated challenges. Ultrason Sonochem 91:106234
Tiwari BK (2015) Ultrasound: a clean, green extraction technology. TrAC Trends Anal Chem 71:100–109

Trigueros E, Sanz MT, Alonso-Riaño P, Beltrán S, Ramos C, Melgosa R (2021) Recovery of the protein fraction with high antioxidant activity from red seaweed industrial solid residue after agar extraction by subcritical water treatment. J Appl Phycol 36:1–14

Ummat V, Sivagnanam SP, Rajauria G, O'Donnell C, Tiwari BK (2021) Advances in pretreatment techniques and green extraction technologies for bioactive from seaweeds. Trends Food Sci Technol 110:90–106

Vásquez V, Martínez R, Bernal C (2019) Enzyme-assisted extraction of proteins from the seaweeds *Macrocystis pyrifera* and *Chondracanthus chamissoi*: characterization of the extracts and their bioactive potential. J Appl Phycol 31:1999–2010

Vázquez-Delfín E, Robledo D, Freile-Pelegrín Y (2014) Microwave-assisted extraction of the carrageenan from Hypnea musciformis (Cystocloniaceae, Rhodophyta). J Appl Phycol 26(2):901–907

Vijayan SR, Santhiyagu P, Singamuthu M, Kumari Ahila N, Jayaraman R, Ethiraj K (2014) Synthesis and characterization of silver and gold nanoparticles using aqueous extract of seaweed, *Turbinaria conoides*, and their antimicrofouling activity. Sci World J. https://doi.org/10.1155/2014/938272

Vinatoru M, Mason TJ, Calinescu I (2017) Ultrasonically assisted extraction (UAE) and microwave assisted extraction (MAE) of functional compounds from plant materials. TrAC Trends Anal Chem 97:159–178

Viganó J, Zabot GL, Martínez J (2017) Supercritical fluid and pressurized liquid extractions of phytonutrients from passion fruit by-products: economic evaluation of sequential multi-stage and single-stage processes. J Supercrit Fluids 122:88–98

Yadav H, Jain S, Rastamanesh R (2012) Fermentation technology in the development of functional foods for human health: where we should head. Ferment Technol 1:10–13

Yang J, Sophia Q, Niu H, Dalai A, Corscadden K, Zhou N (2020) Microwaveassisted hydrothermal liquefaction of biomass model components and comparison with conventional heating. Fuel 277:118202. https://doi.org/10.1016/j.fuel.2020.118202

Yew YP, Shameli K, Miyake M, Kuwano N, Bt Ahmad Khairudin NB, Bt Mohamad SE, Lee KX (2016) Green synthesis of magnetite (Fe_3O_4) nanoparticles using seaweed (Kappaphycus alvarezii) extract. Nanoscale Res Lett 11(1):1–7. https://doi.org/10.1186/s11671-016-1498-2

Zhang Y, Hawboldt K, MacQuarrie S, Thomas R, Gebregiworgis T (2024) Alkaline subcritical water extraction of bioactive compounds and antioxidants from beach-cast brown algae (*Ascophyllum Nodosum*). Chem Eng J 494:153109

Zhang Q-W, Lin L-G, Ye W-C (2018) Techniques for extraction and isolation of natural products: a comprehensive review. Chin Med 13:20

Zollmann M, Robin A, Prabhu M, Polikovsky M, Gillis A, Greiserman S, Golberg A (2019) Green technology in green macroalgal biorefineries. Phycologia 58:516–534

Chapter 5
Seaweeds: Innovation, Challenges to Sustainability

5.1 Seaweed as an Equitable Blue Economy

The development of the seaweed industry emerges as a crucial element in the transition toward a more inclusive and equitable blue economy. Seaweed farming technologies that are socially acceptable. Environmentally benign and commercially viable should be promoted. National and international authorities in tropical and subtropical countries have developed best management practices. Ensuring that these nourishing aquatic crops contribute to the health of the land and its people should be done as soon as possible. The hardest thing about thinking of seaweed as a daily diet is making seaweed food products cheap, tasty, and accessible to billions of people. Moreover, it is challenging to keep customers from being confused about whether products made with seaweed are nutritionally equivalent to the other meals they eat (Wang et al. 2020, 2025). Any new product (food, drugs, cosmetics, and industrial) needs time to be accepted by society and the world, and in this situation, seaweed is an entirely new dietary component for many regions of the world. As a result, winning consumer acceptance will be tough. (1) To establish a socioeconomic structure that allows for the circulation of products in natural settings. (2) To ensure that humans and the environment coexist peacefully. (3) To foster a society in which all parties take part in environmental preservation efforts. (4) To promote global environmental activities. These objectives can be utilized to conserve and restore the marine environment (Hariram et al. 2023).

G. M. El Zokm et al., *Rethinking Seaweeds via Applications and Technologies*, Environmental Science and Engineering, https://doi.org/10.1007/978-3-032-14956-5_5

5.2 Seaweed (Future Food) to Overcome Sustainability Issues

From a sustainability standpoint, seaweed holds great promise in addressing sustainability issues, including pollution, the depletion of the planet's resources, and the need to develop new foods to meet demand. Consequently, the full life cycle of the application, including use and end-of-life, should be considered in the environmental life cycle assessment of seaweed production and application. It may be necessary to redefine the positive appraisals of the environmental benefits associated with sequestering carbon and nutrients. All the material covered in this book will aid in understanding the pressing need to support the possibility of achieving food security worldwide through massive seaweed farming for climate-resilient mariculture, as well as the need for additional research on a range of seaweed farming-related subjects. Seaweed is already widely used in the food, cosmetic, pharmaceutical, and agriculture industries and has potential as a biofuel. According to a report by the Food and Agriculture Organization (FAO 2024), global seaweed production has surpassed 37.8 million tons annually, with Asia contributing approximately 97% of the total. This chapter solidifies the potential capacity of seaweeds as a vital component in reaching global SDGs (Fig. 5.1).

Global seaweed farming is important for industrial applications, women's empowerment, climate change adaptation and mitigation, and food and nutritional security. Thus, the purpose of this study is to record the role and future potential of seaweed farming in these areas. The following are the primary questions that are addressed here: (1) How may seaweed farming potentially improve the nutritional aspects of global food security? (2) Can the cultivation of seaweed aid in the adaptation and mitigation of the recent global climate change? (3) How does seaweed farming, especially in developing nations, support the resilience of livelihoods and the empowerment of women in vulnerable coastal communities? (4) What problems and difficulties does seaweed farming present in relation to the aforementioned? (Figs. 5.2 and 5.3).

5.3 Sustainability of Seaweed for Economic Prosperity and Social Equity

Enhancing the sustainability of seaweed products should prioritize the use of recyclable materials for farming operations and renewable energy sources for hatchery electricity, which have been highlighted as important means of lowering farming emissions (Pessarrodona et al. 2024).

Numerous researchers evaluated seaweed sustainability via economic viability and social equality (Ismail and El Zokm 2025; Hinge et al. 2025; Jueterbock et al. 2025; Canvin et al. 2025; Bullen et al. 2024; Khan et al. 2024; Trif et al. 2024; Spillias et al. 2023; DeAngelo et al. 2023; Dolliver and O'Connor 2022; Hasselström et al. 2020; Hauschild 2015; ISO 2012). Effectively managing residual aquatic seaweed

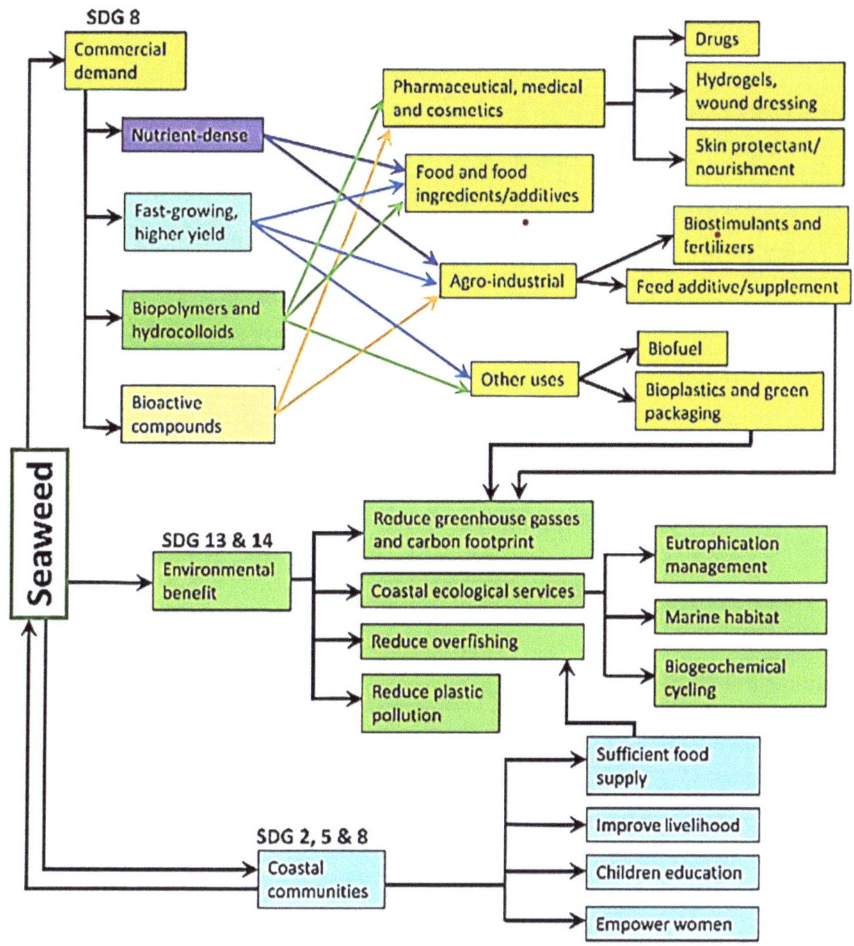

Fig. 5.1 The complex interconnections between commercially sought seaweed characteristics and their possible benefits to the United Nations' Sustainable Development Goals (SDGs) (Sugumaran et al. 2022)

biomass in a sustainable and cost-efficient manner necessitates a thorough approach that takes into account environmental factors as well as economic viability (Trif et al. 2024). The use of residual aquatic seaweed biomass is a dynamic research field, with specific applications and technology likely to advance as additional studies are undertaken.

Seaweed can improve nutrition, promote sustainable agriculture, and guarantee food security. Due to reduced agricultural water supplies, less fertile land, and lower traditional farming earnings, seaweed production has emerged as a lucrative industry in many nations. Seaweeds are a sustainable choice because they don't need fresh

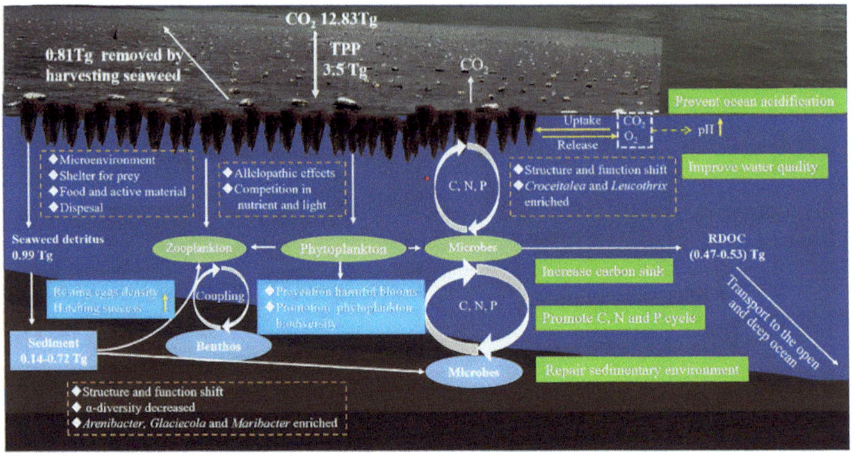

Fig. 5.2 An outline of the main ecological processes that make up the ecology of seaweed farming. RDOC stands for refractory dissolved organic carbon, and TPP for total primary productivity (Wang et al. 2025)

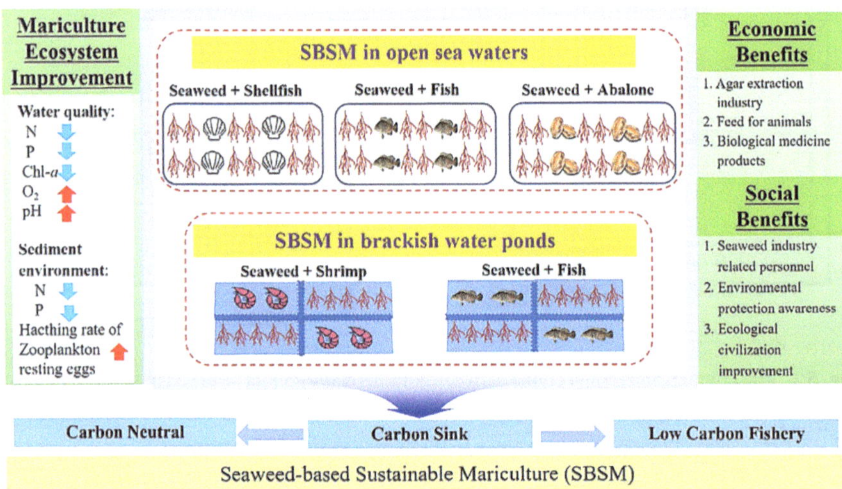

Fig. 5.3 An overview of seaweed-based sustainable mariculture (SBSM) and its advantages for the economy, society, and environment (Wang et al. 2025)

water or outside nutrients to flourish. For cultivation to be effective, regular care is necessary, even though a small amount of labor is needed.

The establishment of a sustainable seaweed farming industry requires improved cooperation and coordination between nations that produce and consume seaweed (#**SDG17**).

The enduring influence of both natural and anthropogenic factors on the system constitutes a facet of sustainability. One of the markers of sustainability, encompassing social equality, environmental considerations, and economic factors, is "**Fisher Information Theory**." It can be characterized in three ways:

$$\text{Eco-efficiency} = \frac{\text{Functionality provided}}{\text{Environmental impact}}$$

Two sustainability indicators were developed:

(i) Net CO_2 balance and energy returned on energy invested (EROEI).
(ii) The energy produced and consumed in each scenario was assessed to determine the EROI.

Quality, bioresource availability, logistic planning, policy, research, and innovation, together with economic, ecological, and social issues, are the main categories of bottlenecks identified in seaweed-based supply chains. These components are discussed in more detail in the following sections, which also examine how new perspectives and approaches could aid in removing obstacles and integrating the social component of sustainability (Valero et al. 2018).

5.4 The Potential for Seaweeds to Play a Greater Role in Mitigating Climate Change

It is currently a contentious topic (Hurd et al. 2022; Luo et al. 2024; Pessarrodona et al. 2024). There are four suggested methods for utilizing seaweed in efforts to mitigate climate change (Ross et al. 2023).

(1) Maintaining and restoring natural seaweed ecosystems.
(2) Growing the use of sustainable nearshore seaweed farming.
(3) Reducing industrial CO_2 emissions with seaweed products.
(4) Immersing seaweed in the deep sea to capture CO_2.

Measuring the total effect of carbon export from seaweed farming and restoration sites on atmospheric CO_2 is still fraught with relevant uncertainty. The scalability of this process is still unclear, although there is evidence that nearshore seaweed farming helps store carbon in the sediments beneath farm locations (Ross et al. 2023). Although using seaweed mariculture products, such as low-carbon food resources or the methane-reducing seaweed *Asparagopsis*, to combat climate change has potential, their carbon footprint and emission reduction potential have not been thoroughly evaluated (Wang et al. 2025). Furthermore, considering the lack of knowledge regarding seaweed biomass's ability to mitigate climate change, purposefully cultivating and immersing it in the open ocean has environmental problems (Ross et al. 2023).

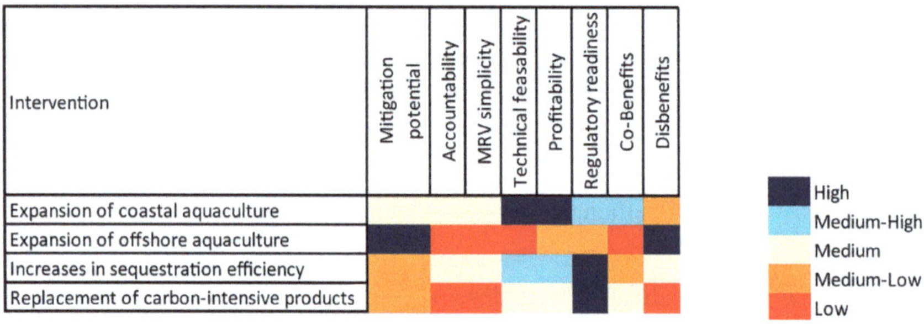

Fig. 5.4 Evaluation of seaweed farming practices that may have a role in climate change (Pessarrodona et al. 2024)

The ocean is a key component of the portfolio of solutions for carbon dioxide removal (CDR) from the atmosphere, which is being developed in response to the urgent need to lessen the consequences of climate change. The government and commercial sectors are beginning to view the growth of seaweed farming as one of the possible CDR options. However, there are still few thorough studies looking at whether farming can actually reduce the effects of climate change. The majority of the carbon fixed by seaweeds is stored in short-term carbon reservoirs, so here we review the findings of more than 100 publications to summarize the evidence about the CDR capacity of seaweed farms and the various interventions that could help mitigate climate change.

By replacing more carbon-intensive items, seaweed products derived from the harvested biomass may help mitigate climate change and save greenhouse gas emissions (Pessarrodona et al. 2024) (Fig. 5.4).

The computations shown here and those found in the literature serve as the basis for scores. Every criterion refers to the state of knowledge at the time. The term "mitigation potential" describes the total amount of possible mitigation. "Accountability" refers to how well interventions operate with the frameworks and methods now used for carbon accounting. "MRV simplicity" describes how difficult it is to track, record, and validate each intervention's GHG fluxes in order to produce credible mitigation. The viability of scaling up the intervention is referred to as "technical feasibility." Potential revenues based on current market pricing are referred to as "profitability." "Regulatory readiness" describes the difficulty and expense of securing agricultural permits in both domestic and foreign countries. "Co-benefits" are any benefits that come from the intervention that are not mitigation (e.g., biodiversity, social uplift). 'Disbenefits' refers to any potential negative impacts resulting from the intervention.

By absorbing between 25 and 50% of the total anthropogenic emissions per year, the world's oceans play a crucial role in controlling the quantity of CO_2 in the atmosphere. Strong photosynthesis and high productivity are characteristics of marine macro-aquatic plants. Because of its capacity to sequester carbon and its ecological significance in reducing the eutrophication of water bodies, seaweed, in particular, has attracted a lot of attention (Kwan et al. 2022). Determining the carbon source

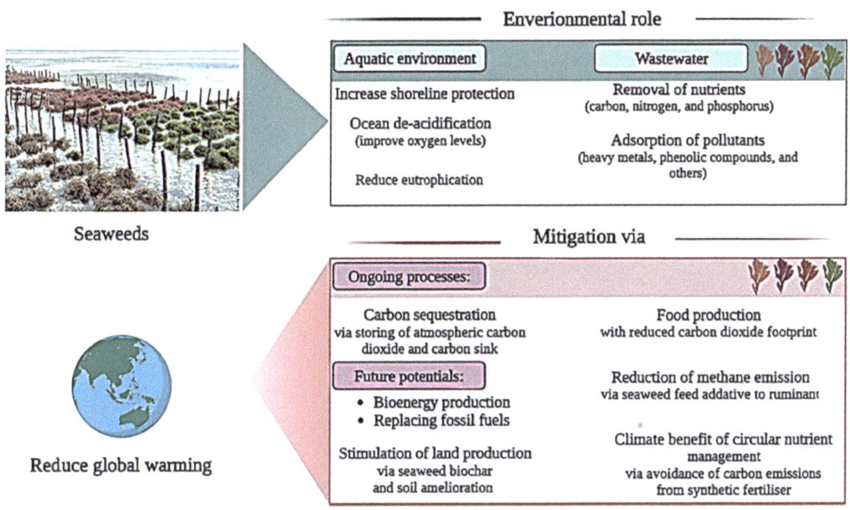

Fig. 5.5 Seaweeds' beneficial roles in reducing climate change and restoring the environment (Farghali et al. 2023)

or sink of crops at various phases of growth within the agricultural environment is difficult. Plants have a limited capacity to store carbon, and the majority of them will return carbon to the environment. Seaweed has a growth cycle that lasts around a year and shows comparatively high photosynthesis and production.

Seaweed that grows on the continental shelf has a carbon sink capacity equal to 35% of the total oceanic carbon sequestration, and certain algae may remove inorganic carbon (IC) with an efficiency of up to 96.78%. Additionally, seaweed can absorb excess nutrients from the environment, which helps to maintain, improve, and restore the marine ecological environment (Fig. 5.5), (Wang et al. 2025).

5.5 Challenges to Seaweed Growth

One of the biggest threats to humans and the world's ecosystems today is climate change, which is mostly caused by rising atmospheric carbon dioxide (CO_2). In order to reach global climate targets, carbon dioxide removal (CDR) technologies are becoming more and more important, and seaweed farming has recently drawn attention as a promising strategy. Interest in potential biomass-based negative emissions and bioenergy, notably from marine sources like seaweed, is being fueled by net-zero greenhouse gas (GHG) emissions targets. However, a thorough evaluation of the economic and biophysical constraints on seaweed farming at scales pertinent to the global carbon budget has not yet been conducted (Ricart et al. 2022; DeAngelo et al. 2023; Ross et al. 2023).

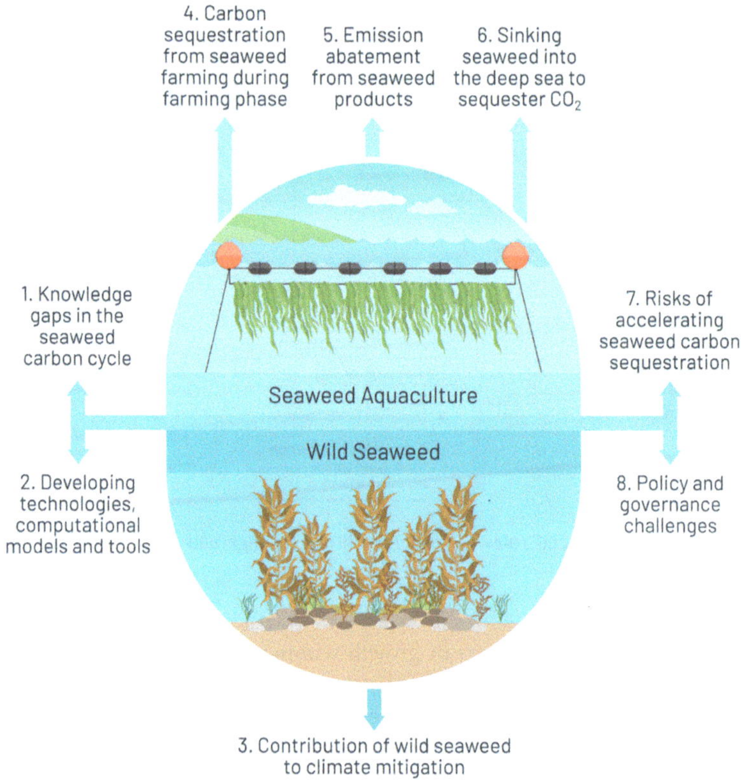

Fig. 5.6 A graphic shows how the stated research problems relate to either seaweed farming or wild/naturally occurring seaweeds. The research challenges have been shortened (Ross et al. 2023)

Figure 5.6 lists eight major legislative and scientific obstacles to the feasibility of seaweed programs for mitigating climate change. Seaweed projects' potential for mitigating climate change is examined, together with the current state of knowledge and important gaps pertaining to these eight major issues (Ross et al. 2023). Seaweed will continue to be seen as one of the possible options for mitigating the effects of climate change, as governments and industries are under increasing pressure to develop mitigation techniques. Seaweed carbon accounting is still in its infancy, albeit (Hurd et al. 2022).

5.6 Future Prospective (Proposed Scenarios)

Global Biosphere Management Model (GLOBIOM)

There are several global scenarios of greater use of seaweed for fuel, feed, or food that would signify a significant change in each of their respective industries. In comparison to a baseline trajectory where raw material demand is satisfied without seaweed, we estimate the environmental benefits of each scenario by 2050 with respect to variations in land-use change, GHG emissions, water and fertilizer use, and species presence changes using the Global Biosphere Management Model (GLOBIOM). Lastly, we employ a spatial prioritization algorithm to determine the most likely locations for seaweed farming and, consequently, the amount of sea used to satisfy demand in each scenario. The model is designed to project how seaweed farming would affect emissions, water and fertilizer consumption, and land use in 2050. Each scenario was compared to a baseline scenario (Table 5.1) (Spillias et al. 2023).

Future studies should assess how successfully the roadmap balances social, economic, and environmental aspects to promote sustainable seaweed farming. To evaluate the practical effects of these sustainable behaviors, longer-term research is required.

Table 5.1 The impact of seaweed replacement on the environmental effects of land use and the accompanying rise in sea-use change was estimated using five seaweed demand scenarios (Spillias et al. 2023)

1-Food	• Seaweeds represent 10% of global diets through energy content by 2050.
2-Feed	• Seaweeds reveealed 10% of global livestock feed by energy content by 2050.
3-Fuel	• Seaweeds have 50% of the feedstock used to produce transport fuels by 2050.
4-All	• All Scenarios 'Food', 'Feed', and 'Fuel' are all realized by 2050.
5-Aspa	• *Asparagopsis* supplements 0.5% of all ruminant diets by 2050, reduces methane production per animal by 68% and increases food conversion efficiency by 14%25

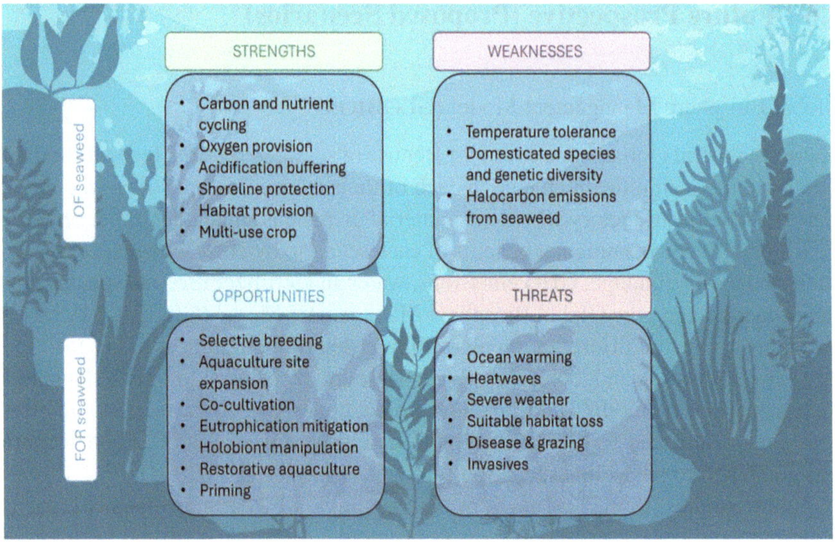

Fig. 5.7 An overview of the benefits and drawbacks of seaweed farming in light of climate change, as well as the potential and difficulties the seaweed sector faces (Veenhof et al. 2024)

5.7 A Look at Potential Future Innovations that Could Influence the Direction of the Seaweed-Promising Business

Seaweed production has environmental, economic, and food security advantages and has the potential to significantly impact the future. However, for this to be actualized, a balance must be struck between economic feasibility and competitiveness within the current economy. Opportunities for increasing resilience and promoting sustainable development in the seaweed industry are illustrated in Figs. 5.7 and 5.8.

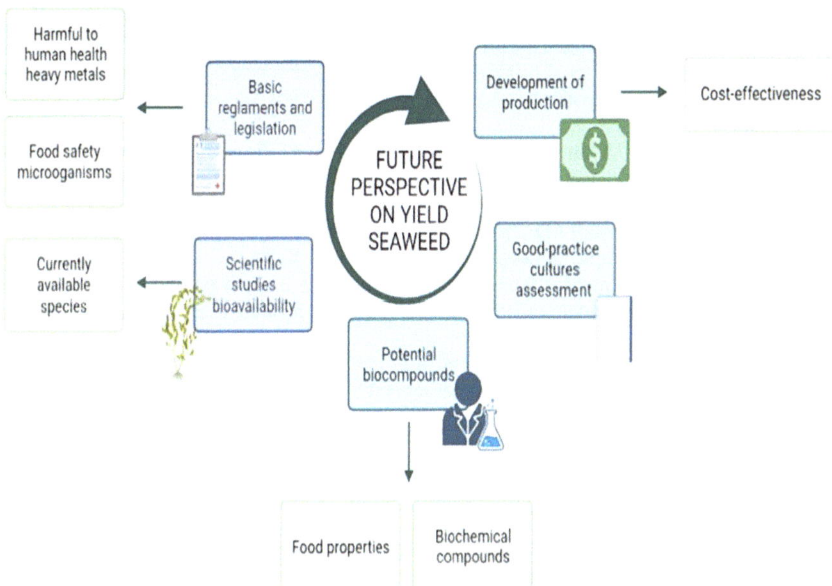

Fig. 5.8 An overview of the main ideas in seaweed development (Salido et al. 2024)

References

Bullen CD, Driscoll J, Burt J et al (2024) The potential climate benefits of seaweed farming in temperate waters. Sci Rep 14:15021. https://doi.org/10.1038/s41598-024-65408-3

Canvin MC, Borrero-Santiago AR, Brook T, Gupta M, Knoop J, Menage G, Moore PJ, O'Connor NE, Ricart AM, Smale DA (2025) Can the emerging European seaweed industry contribute to climate change mitigation by enhancing carbon sequestration? Rev Aquac 17(2):e70004. https://doi.org/10.1111/raq.70004

DeAngelo J, Saenz BT, Arzeno-Soltero IB et al (2023) Economic and biophysical limits to seaweed farming for climate change mitigation. Nat Plants 9:45–57. https://doi.org/10.1038/s41477-022-01305-9

Dolliver J, O'Connor NE (2022) Estimating growth, loss and potential carbon sequestration of farmed kelp: a case study of Saccharina latissima at Strangford Lough, Northern Ireland. Appl Phycol 3(1):324–339

FAO (2024) The state of world fisheries and aquaculture 2024. FAO. https://doi.org/10.4060/cd0683en

Farghali M, Mohamed IMA, Osman AI et al (2023) Seaweed for climate mitigation, wastewater treatment, bioenergy, bioplastic, biochar, food, pharmaceuticals, and cosmetics: a review. Environ Chem Lett 21:97–152. https://doi.org/10.1007/s10311-022-01520-y

Hariram NP, Mekha KB, Suganthan V, Sudhakar K (2023) Sustainalism: an integrated socio-economic-environmental model to address sustainable development and sustainability. Sustainability 15:10682. https://doi.org/10.3390/su151310682

Hasselström L, Thomas JB, Nordström J et al (2020) Socioeconomic prospects of a seaweed bioeconomy in Sweden. Sci Rep 10:1610. https://doi.org/10.1038/s41598-020-58389-6

Hauschild MZ (2015) Better, but is it good enough? On the need to consider both eco-efficiency and eco-effectiveness to gauge industrial sustainability. Procedia CIRP 29:1–7

Hinge M, Grilo VA, Jehn FU et al (2025) Seaweed cultivation: a cost-effective strategy for food production in a global catastrophe. Aquacult Int 33:344. https://doi.org/10.1007/s10499-025-01978-x

Hurd CL, Law CS, Bach LT, Britton D, Hovenden M, Paine E, Raven JA, Tamsitt V, Boyd PW (2022) Forensic carbon accounting: assessing the role of seaweeds for carbon sequestration. J Phycol 58(3):347–363

Ismail MM, El Zokm GM (2025) Algae as keystone for blue economy: sustainability and challenges. Discov Sustain 6:366. https://doi.org/10.1007/s43621-024-00746-w

ISO (2012) Environmental management—eco-efficiency assessment of product systems—principles, requirements and guidelines (ISO 14045:2012). International Organization for Standardization, Geneva

Jueterbock A, Hoarau-Heemstra H, Wigger K et al (2025) Roadmap to sustainably develop the European seaweed industry. npj Ocean Sustain4:22. https://doi.org/10.1038/s44183-025-001 22-9

Khan N, Sudhakar K, Mamat R (2024) Macroalgae farming for a sustainable future: navigating opportunities and driving innovation. Heliyon 10:e28208. https://doi.org/10.1016/j.heliyon.2024.e28208

Kwan V, Fong J, Ng CSL, Huang D (2022) Temporal and spatial dynamics of tropical macroalgal contributions to blue carbon. Sci Total Environ 828:154369

Luo H, Li J, Xie S, Yang Y (2024) Understanding and estimating the role of large-scale seaweed cultivation for carbon sequestration on a global scale over the past two decades. Sci Total Environ 955:176966

Pessarrodona A, Howard J, Pidgeon E, Wernberg T, Filbee-Dexter K (2024) Carbon removal and climate change mitigation by seaweed farming: a state of knowledge review. Sci Total Environ 918:170525. https://doi.org/10.1016/j.scitotenv.2024.170525

Ricart AM, Krause-Jensen D, Hancke K, Price NN, Masqué P, Duarte CM (2022) Sinking seaweed in the deep ocean for carbon neutrality is ahead of science and beyond the ethics. Environ Res Lett 17(8):81003. https://doi.org/10.1088/1748-9326/ac82ff

Ross FWR, Boyd PW, Filbee-Dexter K, Watanabe K, Ortega A, Krause-Jensen D et al (2023) Potential role of seaweeds in climate change mitigation. Sci Total Environ 885:163699. https://doi.org/10.1016/J.SCITOTENV.2023.163699

Salido M, Soto M, Seoane S (2024) Seaweed: nutritional and gastronomic perspective. A reivew. Algal Res 77:103357. https://doi.org/10.1016/j.algal.2023.103357

Spillias S, Valin H, Batka M et al (2023) Reducing global land-use pressures with seaweed farming. Nat Sustain 6:380–390. https://doi.org/10.1038/s41893-022-01043-y

Sugumaran R, Padam BS, Yong WTL, Saallah S, Ahmed K, Yusof NA (2022) A retrospective review of global commercial seaweed production—current challenges, biosecurity and mitigation measures and prospects. Int J Environ Res Public Health 19:7087. https://doi.org/10.3390/ijerph19127087

Trif M, Rusu AV, Bellahcen TO, Cherifi O, El Bakali M (2024) Sustainable and cost-effective management of residual aquatic seaweed biomass. Business opportunity for seaweeds biorifineries. In: Ozogul F, Trif M, Rusu A (eds) Seaweeds and seaweed-derived compounds. Springer, Cham. https://doi.org/10.1007/978-3-031-65529-6_12

Valero A, Valero A, Calvo G, Ortego A (2018) Material bottlenecks in the future development of green technologies. Renew Sustain Energy Rev 93:178–200

Veenhof RJ, Burrows MT, Hughes AD, Michalek K, Ross ME, Thomson AI, Fedenko J, Stanley MS (2024) Sustainable seaweed aquaculture and climate change in the North Atlantic: challenges and opportunities. Front Mar Sci 11:1483330. https://doi.org/10.3389/fmars.2024.1483330

Wang Q, Sun X, Lin S, Dong Y, Shen H, He Z, Luo H, Zou L, Chung IK, Yang Y (2025) Large-scale seaweed cultivation as a nature solution for a carbon-negative economy and restorative environmental stewardship: lessons, 2025 from China. Renew Sustain Energy Rev 207. https://doi.org/10.1016/j.rser.2024.114954

Wang X, He L, Ma Y, Huan L, Wang Y, Xia B, Wang G (2020) Economically important red algae resources along the Chinese coast: history, status, and prospects for their utilization. Algal Res 46:101817

Conclusion

The enormous potential of macroalgae farming to combat climate change and accomplish the Sustainable Development Goals of the UN has been examined in this chapter. By outlining the current state of seaweed production worldwide, different growth techniques, current trends, and difficulties, a thorough grasp of this expanding seaweed sector is given. To fully realize the advantages that macroalgae farming can provide on a global scale, certain basic issues must be resolved. The study's main conclusions are as follows.

1. Seaweed Farming Innovations: The development of cutting-edge, cost-effective, and environmentally friendly methods is crucial to meeting the growing demand for macroalgae production. It is necessary to develop sustainable farming practices that strike a balance between environmental protection and output.

For the industry to prosper in a changing climate, adaptive farming techniques must also be investigated. Countries may contribute to sustainable seaweed production globally and realize the enormous potential of macroalgae agriculture by putting appropriate measures into place.

2. Addressing the Obstacles: Despite significant prospects, seaweed cultivation encounters numerous problems. It is essential to meticulously choose appropriate production locations to alleviate issues stemming from climate change, epiphytes, and the proliferation of pathogenic species. Biotic and abiotic variables can profoundly influence output, necessitating ongoing research and innovation. Creating disease-resistant varieties, enhancing production techniques, and implementing effective mitigation strategies are crucial for addressing these challenges. Overcoming these obstacles requires the development of disease-resistant breeds, enhanced cultivation technologies, and efficient mitigation techniques. By creating germplasm banks and disease-resistant macroalgae, genetic diversity can be preserved, and sustainable growth can be guaranteed.
3. Seaweed communities can be empowered by providing well-coordinated public funding to macroalgae growers, which will help them produce value and even

G. M. El Zokm et al., *Rethinking Seaweeds via Applications and Technologies*,
Environmental Science and Engineering, https://doi.org/10.1007/978-3-032-14956-5

commercialize their products. Farmers must have access to educational programs that teach them how to manage risks and challenges while also helping them get planting materials and farming equipment at discounted or subsidized costs.

4. Establishing National Standards and Goals: For the purpose of keeping an eye on disease outbreaks and insect reproduction, governments should implement internationally recognized standards and procedures. Adopting uniform procedures across countries will reduce the possibility of biosecurity hazards and encourage prudent resource management. Reaching a specified output level to satisfy market demands while promoting carbon neutrality through seaweed cultivation is the nation-specific goal. Every country can support a thriving and sustainable global seaweed industry by setting clear goals and keeping track of its progress.
5. Cooperation and Knowledge Exchange: To fully realize the promise of seaweed production for a resilient and sustainable future, cooperation between scientists, researchers, legislators, and industry players is essential. Urge investors to encourage seaweed innovation, legislators to enact laws that will help, and scientists to focus their studies on certain issues. We can overcome the obstacles and fully utilize this adaptable and sustainable resource by exchanging best practices, resources, and information. Essentially, macroalgae farming is situated at the intersection of opportunity and difficulty. Focused efforts, creative solutions, and a dedication to sustainability are needed to overcome the obstacles.

The future of a cooperative ocean where seaweed flourishes can be guided by the insights and viewpoints provided by this study, empowering communities, reducing climate change, and opening the door to a better future. As long as there is a market for the products, seaweed research and development will keep growing. Due to established basic knowledge and a lack of practical application for economic activities, many research disciplines may develop slowly. Seaweed can address a number of fundamental human needs and concerns, such as those related to food, energy, health, and the environment. To propel the paradigm shift toward the global seaweed revolution, however, the stakeholders—government, industry participants, farmers, scientists, and consumers—must cooperate.

A Call to Action

Several immediate steps are needed to manage the dangers associated with the increasing push for seaweed as a CDR strategy:

- Increase scientific knowledge about seaweed's use as a CDR method, including its length and scope, as well as the ecological effects that go along with it. the creation of thorough assessments of carrying capacity for large-scale seaweed farming that are specific to a certain region.
- Produce reliable approximations of the spatial area required for sufficient seaweed growth.

- Develop ethical and sustainable guidelines for farmed seaweed or large-scale natural nuisances, taking into account all relevant UN SDGs and accounting for socioeconomic and climate advantages.
- Compile life cycle assessments (LCAs) to take the overall amount of greenhouse gas emissions into consideration.
- Take into account every alternative that seaweeds might provide as a natural way to lower CO_2 emissions. One step in incorporating all seaweed uses (like food and biofuel) and their distinct temporal scale of carbon sequestration in estimates of carbon neutrality is to consider the role of all carbon pathways from seaweed cultivation, including lost particle organic carbon and excreted dissolved organic carbon, which are usually ignored.